To Dad,
Hope you find this book informative on God's Creation. The Chapter on Cosmic History should be interesting!
Merry Christmas '92.
We love you,
Patty, Mike
Sarah + Kathleen

PORTRAITS OF CREATION

PORTRAITS OF CREATION

Biblical and Scientific Perspectives on the World's Formation

Howard J. Van Till, Robert E. Snow
John H. Stek, Davis A. Young

CALVIN CENTER FOR CHRISTIAN SCHOLARSHIP

WILLIAM B. EERDMANS PUBLISHING COMPANY
GRAND RAPIDS, MICHIGAN

255 Jefferson Ave. S.E., Grand Rapids, Mich. 49503

Printed in the United States of America

Library of Congress Cataloging-in-Publication Data

Portraits of creation : biblical and scientific perspectives on the world's formation / Howard J. Van Till . . . [et al.].

p. cm.

ISBN 0-8028-0485-3

1. Creation. 2. Creationism. I. Van Till, Howard J., 1938- .

BS651.P67 1990

231.7'65—dc20 90-35562

CIP

CONTENTS

ACKNOWLEDGMENTS

THIS BOOK is a product of research and writing begun while the authors were Fellows of the Calvin Center for Christian Scholarship (CCCS) of Calvin College, Grand Rapids, Michigan. We are deeply grateful for that opportunity to work as a community of Christian scholars, and we hereby express our gratitude not only to the governing board of the CCCS but also to the many people whose financial support has made possible this continuing program of interdisciplinary Christian scholarship.

With each chapter of this work we have provided the name of its principal author. But this work was truly a *team* effort, so that each of us must acknowledge the benefit of the assistance and constructive criticism contributed by the other members of this research team, which included not only the authors of this book but also Prof. George M. Marsden, now at the Divinity School, Duke University; Dr. Clarence Menninga, professor of geology, Calvin College; and the Rev. John Suk, who served while a student at Calvin Theological Seminary. Chapter 8 was written while its author (H.V.T.) was in residence at the Center of Theological Inquiry at Princeton, New Jersey, and we thank the Center for its provision of helpful resources and a stimulating environment for reflecting on the place of this work in a much larger and ongoing discussion.

Finally, a special word of thanks to Tim Straayer of William B. Eerdmans Publishing Company for his skillful assistance in taking this book from manuscript to finished product.

Grand Rapids
January 1990

Howard J. Van Till, Robert A. Snow,
John H. Stek, Davis A. Young

INTRODUCTORY COMMENTS BY THE COORDINATOR

AS THE EDITOR of this collaborative volume and as the Coordinator of the CCCS study team on the topic "Creation and Cosmogony," I would like to provide some background information and a few personal comments that may assist the reader in understanding the scope both of our particular agenda and of what remains to be investigated.

The original proposal for this interdisciplinary study included, in addition to the topics addressed in this volume, a concern to address questions in the arenas of evolutionary biology and anthropology. Both of these fields of study were identified as deserving a careful critique from the viewpoint of a firm commitment to the historic Christian faith. But the CCCS governing board judged, and this team concurred in that judgment, that such important topics should not be crowded into an already heavy agenda but should be the focus of investigation by a research team specifically selected for that purpose at some future time. Continuing and highly varied expressions of concern for these matters, especially by anxious and sometimes strident critics of the scientific enterprise, strongly affirm our judgment that further study is urgently needed.

We hope that the product of our research and interdisciplinary dialogue will contribute fruitful insights on many questions regarding the relationship of biblical, theological, and scientific scholarship, but we candidly recognize not only the omission of other important questions but also that our work will itself raise some questions that are not fully resolved in this volume. Such is the nature of scholarship, and our omission of a particular relevant question should never be construed to imply either that we have carelessly overlooked it or that we are purposely withholding our perspective on it; rather, it should simply be seen as evidence that we were not able to treat all such questions within the tenure of our study and had to limit our agenda in numerous ways.

Several questions of this sort could be cited. Consider, for example, questions arising out of the diversity of concepts held by Christians regarding the "historicity" or the "event character" of the stories narrated in Genesis 1–11. If the focus of Genesis 1 is *theological* rather than *chronological* (see our chap. 7)—that is, if this narrative was written in an artful form endemic to ancient Near Eastern culture and provided by God to serve as a basis for theological doctrine but not to function as a chronicle of historical particulars with which our modern scientific theorizing must be forced to concur (see our chap. 4), and if three centuries of geological investigation convincingly precludes an interpretation of Genesis 6–9 in terms of a planet-encompassing flood that killed nearly all creatures and all but eight humans (see our chap. 3)—then what do we Christians mean when we affirm the "historicity" or the "event character" of Genesis 1–11, especially of chapters 2 and 3? Granted our firm conviction that Genesis 1–11 is concerned with authentic history, are we also demanding that it provide us with historical particulars of the kind expected in our modern Western culture, or are we instead affirming more generalized statements regarding the revelational meaning of these narratives and their theological implications regarding the historical authenticity of God's directing the formation of "the heavens and the earth" and the development of humankind from "the dust of the earth"?

And what are we to think of the concept of biological evolution, properly limited to those matters on which the natural sciences are competent to speak (see our chap. 5), including the evolution of humankind as special and morally responsible creatures? Does the continuity and consistently patterned activity perceived in the formative history of inanimate systems (like planets, stars, and galaxies) provide sufficient warrant for expecting genealogical continuity in the formation of all living creatures as well? Even of humankind? Or do we have biblically informed theological grounds for excluding that possibility? What if the concept of macroevolution continues to gain credibility from empirical evidence? How might concepts drawn from scientific anthropology be permitted to influence our understanding or articulation of theological anthropology? And if we have for decades neglected to pay due attention to the results of physical and cultural anthropology, do we have any right to expect that statements of theological anthropology articulated primarily in the conceptual vocabulary of the sixteenth and seventeenth centuries will be adequate for addressing questions raised by twentieth-century Western culture and occasioned in part by the knowledge gained through contemporary empirical science?

Other specific questions could also be cited as issues of continuing concern, even anxiety, in the arena of the relationship between empirically informed natural science and biblically informed Christian theology. But are there some general features that tie these questions together—features that may shape the continuing investigative efforts that we heartily encourage? In my judgment there are; let me briefly explain.

As Christians we rightly seek to grow in our understanding of God and of his works in this world. Focusing our attention here on our efforts to understand the physical universe as the Creation with which God continues to interact, we desire to construct and evaluate our theoretical models by drawing upon all that we know about the world. Specifically we wish to incorporate both what we know (or think that we know) by empirical study of the created world and what we know (or think that we know) by exegetical study of the Scriptures.

In drawing from each resource, however, we face difficult questions regarding both epistemology and hermeneutics. What do we really *know,* for example, from the results of empirical science? In the context of conflicting claims by "experts," some genuine, some bogus, how do we come to *know* that a particular scientific theory should be held with a high degree of confidence? Have the relevant empirical data been competently *interpreted?* Have we made the move from "raw data" to systematizing and interpretive theory with the requisite level of scientific proficiency and integrity (see our chap. 5)? And in making this move, have we made full and proper use of what we know from a faithful and well-informed study of the Scriptures?

Likewise, what do we really *know* from the results of biblical exegesis? In the context of conflicting claims by "experts," some genuine, some bogus, how do we come to *know* that a particular theological theory should be held with a high degree of confidence? Have the relevant biblical data been competently *interpreted?* Have we made the move from "raw text" to systematizing and interpretive doctrine with the requisite level of exegetical proficiency and integrity? And in making this move have we made full and proper use of what we know from an honest and skillful investigation of the Creation?

In my personal judgment the majority of contemporary disagreements among Christians concerning the relationship between natural science and Christian belief arise not from differences in the degree of Christian faithfulness or theological orthodoxy among the contenders but rather from substantial differences in the epistemology and hermeneutics being employed in arriving at positions with respect to empirical science and biblical interpretation. Hence the resolution of these disagreements will be achieved neither by heresy trials in the ecclesiastical court nor by those spirited contests of polemical rhetoric in the spectator sport known as "creation-evolution debates." Resolution will become possible only when, in the atmosphere of mutual respect, Christian brothers and sisters sit down and diligently do their homework together, drawing from the rich resources of both biblical and scientific scholarship (see our chap. 8).

We make no pretense of having fully resolved the issues cited above. Hence we urgently encourage further interdisciplinary investigation on these important questions—not because our commitment to God in faith is dependent on the knowledge sought, but because, having committed our hearts to God, we are called to serve him not only with all of our heart and body but also *with all of our mind.* And this we would do as an expression of our confidence that whatever authentic knowledge we gain will serve only to increase our faith in God, our Creator, Redeemer, and Comforter, and to enhance our experience of his ceaseless work in the world that is here only because he has said and continues to say, "Let there be Creation."

Howard J. Van Till
Professor of Physics
Calvin College

1. WHERE ARE WE?

Perceived Tensions between Biblical and Scientific Cosmogonies

DAVIS A. YOUNG

THE INFLUENCE OF SCIENCE ON CHRISTIANITY

ADVANCES IN SCIENTIFIC KNOWLEDGE of the world have undeniably made an enormous impact on Christian thought and life. They have even affected how Christians read the Bible. For example, in Psalm 24:1-2 we read that "The earth is the LORD's, and everything in it, the world, and all who live in it; for he founded it upon the seas and established it upon the waters."[1] When interpreted in a woodenly literalistic manner, this text appears to claim that the earth rests upon water.[2] Indeed, many Christians, especially in the seventeenth century, concluded that the solid land on which human beings dwell rests upon a vast subterranean ocean or abyss.[3] This belief in vast reservoirs of water beneath the earth was reinforced by the reference in Genesis 7:11 to the fountains of the great deep as one contributor to the Noahic flood.

Other Christians, however, have allowed the science of their day to influence their reading of Psalm 24:2. For example, those who lived in times when Aristotelian science was dominant were convinced that the "natural place" for solid earth was below that of water. They considered it improbable that the earth could exist suspended above a large body of

1. All biblical quotations in this chapter are from the New International Version.

2. A great deal of evidence, both biblical and extrabiblical, witnesses to the fact that people throughout the ancient Near East shared the view that the visible world (i.e., heaven and earth) was bracketed by an ocean of water above and an ocean of water below. See also Ps. 42:7. For further discussion, see chapter 7, pp. 226-29.

3. For example, Thomas Burnet, Robert Hooke, and John Woodward. For more detail on these beliefs, see chapter 3 and also Davis A. Young, "Scripture in the Hands of Geologists (Part One)," *Westminster Theological Journal* 49 (1987): 1-34.

subterranean water. The science of our own day concurs, but for quite different reasons. All manner of theoretical considerations and geophysical experimentation have made it very clear that while there is substantial groundwater in the pores and fissures in rocks, there is no subterranean ocean.

Very few Christians today read Psalm 24:2 literalistically, because they know from modern empirical study that the earth does not rest on water. Yet we hear no complaints that we have not taken God "at his word," or that we have denied biblical authority by allowing science to influence our interpretation of Psalm 24:2.

In speaking of the spread of the gospel, the apostle Paul quoted from Psalm 19:4: "Their voice has gone out into all the earth, their words to the ends of the world" (Rom. 10:18). Some early church fathers used this text to argue against the notion of antipodes—that is, human beings living on the "bottom" side of a spherical planet Earth. They knew that the missionaries of Paul's day had not carried the good news beyond the Mediterranean basin, much less to the alleged other side of the globe. And since Paul stated that gospel preaching had reached the whole world, it was safe to conclude that there could not be any people on the other side of the globe. But the early church fathers who endorsed this reasoning—including the illustrious Augustine—were simply wrong. Centuries of exploration and discovery have convincingly demonstrated that there were long-established civilizations living on the other side of the world from the Mediterranean basin when Paul wrote these words. Armed with this knowledge, Christians no longer read that verse in its most straightforward sense. The reference to "all the earth" is generally interpreted to mean "all of the world known to the apostles at that time."

And then there is Psalm 93:1: "The world is firmly established; it cannot be moved." Read literalistically, this text appears to deny the possibility that the Earth can move. In Galileo's day, this was one of the texts used by ecclesiastical defenders of traditional geocentrism to discount the Copernican hypothesis that the Earth revolved about the Sun. Since then, the newer understanding of solar system mechanics has prevailed even among Bible believers, and Psalm 93:1 has been reinterpreted. Rare is the Christian today who would deny that the Earth spins on its axis once every day, that it revolves around the Sun once every year, or that it is moving along with the rest of the solar system and the galaxy at a staggering pace. Nevertheless, few Christians object that in adopting these views we have wrongfully subjected the Bible to the interpretive influence of scientific discovery.

These are but a small sample of the numerous biblical texts that are no longer read in a strictly literal fashion. The charge that we no longer "take God at his word" would have meaning only if his word were bound in artless slavery to modern scientific standards of verbal precision.[4] The vast majority of Christians "interpret" these texts to mean something other than what they appear literally to say. And our nonliteralistic readings have become so much a part of us that we no longer recognize that we are "interpreting" these texts in the light of acquired knowledge. What is more, these newer interpretations have been developed not only as a consequence of a growing appreciation for the literary artistry of Scripture but also because of the pressures exerted upon readers by scientific discovery. Throughout church history Christians have repeatedly been challenged by scientific discoveries to abandon what were once thought to be the "right" interpretations. Science has frequently provided us with both the occasion and the means to rethink our reading of God's Word.

Science has been appropriated by Christians in other ways as well. Throughout church history, but especially in recent centuries, the natural sciences have often been a valuable ally in the defense of the faith. Christians have not hesitated to appeal to the findings of science to corroborate what was perceived to be the testimony of Scripture. In the seventeenth century, biblical scholars appealed to the current knowledge of the distribution of animals to support the claim that Noah's ark could adequately hold the animal species of the world. In more recent times, Bible commentaries are replete with references to archaeological discoveries that have vindicated or clarified historical statements in Scripture. For example, archaeology has confirmed the existence of the Hittites and other peoples mentioned in the Bible but not in other historical documents.

Nor is it merely in the areas of biblical interpretation and apologetics that Christians have been strongly influenced by science. As much as anyone else in the Western world, contemporary Christians demonstrate by their attitudes and actions that their daily lives have been powerfully affected by the cultural force of modern natural science. One would be hard pressed to find a Western Christian who rejects outright either the discoveries of science or the benefits that come from the

4. See Vern Poythress, *Symphonic Theology* (Grand Rapids: Zondervan, 1987). Poythress stresses that word meanings have "fuzzy boundaries" and warns that demanding overly precise meanings of biblical words may shut us off from important biblical insights. He also warns that we must be careful not to confuse the more precise, technical usage of words in systematic theology with biblical usage.

technology built on that science. Like everyone else, Christians gladly accept the "improvements" that have been made possible by advances in physics, chemistry, and biology. Christians drive cars, watch TV, use computers, and wear pacemakers. Most of us rejoiced at the revelation of new wonders in God's Creation resulting from recent explorations of the Moon and other bodies within the solar system.

THE "TWO BOOKS" METAPHOR

Christians clearly have embraced many of the results of the scientific enterprise. This accepting attitude toward science is not surprising, for there has long been a deep sense within the Christian consciousness that systematic, empirical research into created reality is a valid way in which the nature of God's created world can be discovered. The sixteenth-century Belgic Confession states that

> We know Him by two means: First, by the creation, preservation, and government of the universe; which is before our eyes as a most elegant book, wherein all creatures, great and small, are as so many characters leading us to *see clearly the invisible things of God, even his everlasting power and divinity,* as the apostle Paul says (Rom. 1:20). All which things are sufficient to convince men and leave them without excuse. Second, He makes Himself more clearly and fully known to us by His holy and divine Word, that is to say, as far as is necessary for us to know in this life, to His glory and our salvation.[5]

But from even earlier times Christianity has been powerfully influenced by this "two books" metaphor, which expresses the view that there are two complementary sources of divine revelation: God's written book, the Bible, and God's unwritten book, Creation. Those who hold this view maintain that because the Bible and Creation both come from God, they must be in perfect harmony. What we read in the book of Creation in the context of scientific activity must be in fundamental agreement with our theological reading of the Bible.[6] Christians generally want to accept the

5. Belgic Confession (1561), Art. 2.

6. Without question the "two books" metaphor has proved very fruitful in the history of the relationship between Christian faith and scientific endeavor. But it has also perhaps contributed to some confusion in articulating that relationship. Those who appeal to Creation as a book of God have typically overlooked the emphasis in the Belgic Confession on the fact that the object of its concern is revelation about *God,* not merely information about the created world. The same confusion has frequently arisen over the

scientific enterprise; they see science as a good thing that tells us something about the way God's world works.

THE "WARFARE" METAPHOR

But there has also existed within the Christian community a deep ambivalence toward science. Alongside the positive "two books" metaphor, a negative "conflict" or "warfare" metaphor has also powerfully shaped Christian attitudes toward science.[7] Believers have heard from many Christian pulpits that science and religion are two diametrically opposed ways of looking at the world, that either one is a Christian or one accepts science, but one cannot consistently hold to both. With their suspicions thereby aroused, many Christians have hardened their attitudes toward science into outright hostility when exposed to the blatantly anti-Christian pronouncements of some leading non-Christian practitioners and popularizers of natural science such as Thomas Huxley or Carl Sagan. They have been led to believe that science as presently practiced is fundamentally opposed to Christianity. That belief is further reinforced by the fact that popular television science programs very rarely acknowledge God when presenting the magnificent structure of the galaxy or the delicate balances of nature. Many Christians have become convinced that science is dominated by unbelievers to the point that science itself must have a fundamentally anti-Christian bias.

The "conflict" thesis, however, is most strongly reinforced for its adherents by the impression that science has reached conclusions that contradict the Bible. Such feelings of hostility toward science frequently arise in discussions about the formation and history of the cosmos, of the Earth, and of life. In its investigation of such questions, science has

---

terms "general revelation" and "special revelation." Christians have typically understood "general revelation" as having to do with science. Again, however, the idea is not that *data* are divinely revealed but that *God* is revealed through the created order. Nonetheless, in spite of these qualifications, we are persuaded that both Creation and the Bible are from the same living God and that underlying them both is a fundamental unity that is grounded in God himself.

7. The warfare metaphor began to gain considerable strength during the nineteenth century and was popularized in works on the history of science by John William Draper and Andrew Dixon White. The fact that many clergymen of the nineteenth century were unduly alarmed by developments in science aided in the popularizing of the "military" terminology.

clearly developed conclusions that are at variance with many *traditional* interpretations of the Bible. The Bible has traditionally been read to imply that the universe is young; astronomy concludes that the universe is billions of years old. The Bible has been widely interpreted as saying that the Earth was created in six days; geology concludes that the Earth has undergone a long and complex history spanning 4.5 billion years. The Bible has been interpreted as implying the fixity of animal and plant species; paleontology and biology conclude that organisms have developed from one another through time, that they have evolved. Some believe that the Bible teaches that all death entered the world only after human beings appeared and fell into sin; paleontology concludes that animals and plants died, and in some cases died violently by being devoured by other animals, before human beings were even on the Earth. Many Christians have a very difficult time accepting such conclusions since they cannot see how the Bible can possibly be in accord with them.

Because the majority of Christian practitioners of natural science accept the main conclusions of science within its area of competence, they frequently find themselves the target of the suspicions and even anger of fellow believers. Those who provisionally adopt the findings of science and are willing to explore the possibility of nontraditional interpretations of Scripture are commonly accused of "not taking God at his word." In dealing with the great questions surrounding the beginnings and early history of the world, many Christians object that science is being used to control the interpretation of the Bible. On the other hand, those willing to look for new interpretations are perplexed by the unwillingness of their fellow Christians to consider alternate interpretations of Genesis in response to science since they have already done so with many other texts in Scripture.

It is appropriate to ask why Christians become so agitated over questions of cosmogony. Why do they feel that scientists are tampering with God's Word as soon as they bring geology or astronomy to bear on the early chapters of Genesis? Why is it mainly here that Christian scientists are likely to meet with the objection that they are denying the plain statements of Scripture and are letting science dictate the interpretation of the Bible? History makes it plain that the church has frequently responded to and taken advantage of advances in scientific knowledge. Why then is there such great concern about the formation and history of the cosmos, the Earth, and life? We suggest that at least two main concerns are operative.

CONCERNS ABOUT THE NEW COSMOGONY

First, there exists a latent fear in the hearts of many believers that to depart from a "literal" reading of the early chapters of Genesis is to undermine the fundamental truth that God is the Creator of all things. Concern for this fundamental truth is legitimate, for no biblical teaching is more foundational than the doctrine of creation. And for many Christians the very term *creation* implies the miraculous, virtually instantaneous fulfillment of a divine fiat without the use of existing material or of secondary causes. For them creation *is* the instantaneous production of a mature, fully functioning object out of nothing by God. This view often goes hand in hand with a commitment to the traditional reading of Genesis 1 as involving twenty-four-hour days of creation. With such presuppositions, one would have to conclude that any reading of Genesis 1 or other biblical materials that does not view creation in such terms constitutes a denial of creation *by definition.* And so it is that astronomy and geology, with their talk of ordinary processes operating over billions of years, are perceived as opposed to the doctrine of creation.

Second, many Christians believe that the current scientific cosmogony poses a threat to faith in any divine "intervention" in the course of history in a "supernatural" manner. They frequently express that concern that if science can influence our interpretation of the early chapters of Genesis to the point of removing "miraculous" elements, then other fundamental biblical teachings will also be at risk. What, they ask, is to prevent science from attacking the reality of Jesus' miracles or even his virgin birth and resurrection?[8] If the belief that creation was achieved through a series of supernatural miracles is eliminated, what is to prevent the elimination of all biblical miracles? What is to prevent the substitution of a naturalistic explanation of all unusual elements in the Bible? The fear is that science has the power to bring all the fundamentals of the Christian faith crashing to the ground.[9]

8. Unquestionably many American Christians, vividly remembering the modernist-fundamentalist controversies of the early decades of this century, are afraid of a new turn toward modernism. But those evangelical Christians who are open to new interpretations of the early chapters of Genesis and accept scientific cosmogony have no use for the views expressed by, for example, Harry Emerson Fosdick in *The Modern Use of the Bible* (New York: Macmillan, 1924).

9. The reader must understand that for the authors of this book, these issues are not of purely academic interest. They are for us, too, of intensely personal concern. We greatly empathize with Christians troubled by them, for we too have struggled with them

REACTIONS TO THE NEW COSMOGONY

In response to the perceived threats of an allegedly ungodly science, many Christians have reacted militantly, attacking the modern scientific enterprise in a variety of ways. One of the principal ways has involved the objection that the scientific reconstructions of cosmic, terrestrial, and biological history are merely imaginative speculations without any solid basis. Those who adopt this approach are fond of pointing to the constantly changing content of the theoretical "speculations" concerning cosmic and geological history and biological evolution. They contrast the methodologies of astronomy, geology, and paleontology with those of chemistry and physics. Chemists and physicists, they say, carefully construct their theories on the basis of data that are collected under rigorous laboratory control—data that are in principle reproducible because the relevant experiments can be repeated under identical conditions. In contrast, astronomers, geologists, paleontologists, and scientists in related disciplines deal with past, singular events that are unobservable and unrepeatable. The sort of rigorous laboratory control that is possible in chemistry cannot be achieved in these sciences, they say. How can the theoretical reconstructions of past historical events carry any weight when no one was there to observe them? And so, this reasoning goes, Christians may justifiably dismiss the wild (and probably ungodly) speculations coming out of the astronomy, geology, and biology departments of today's universities, no matter how prestigious they may be.

Another approach to a critique of modern science has been encouraged by developments in Christian apologetics and the philosophy of science. In recent decades Christians have given increasing attention to the very important role that is played by presuppositions and worldviews in the acquisition of knowledge and in the evaluation of scientific theories. Building on the insights of presuppositionalism, many Christians have alleged that modern science has been established on foundations that are essentially hostile to Christianity and the Bible. The principle of uniformity is frequently singled out for criticism on these grounds. Briefly, it is argued that most contemporary interpretation of geological data is incompatible with the biblical worldview because it is

---

personally in many ways and throughout our careers. As much as any other Christians, we have wrestled with the relationship between our professional studies and our personal relationship with God, with Christ, and with the Scriptures.

based on the assumption that the past was fundamentally like the present, whereas the biblical worldview presumably sees the past in terms of a "principle of catastrophism" incompatible with this principle of uniformity.

This point of view has supposedly been reinforced by recent developments in the philosophy of science. In the past, a simplistic view of the nature of science often prevailed in which it was assumed that scientists gathered "facts" or "data" in a purely objective, unbiased manner. However, that view has been replaced by a more honest assessment of the way in which science is practiced. Philosophers of science now recognize that all scientific data collection, theory formulation, and theory evaluation are to some extent "value-laden." There is no purely objective theorizing, and worldviews do influence the way in which data are collected and perceived.

These valid insights have been seized upon by well-meaning believers as a challenge to the conclusions of modern cosmogony. They employ these insights in such a way as to suggest that the data may be interpreted in terms of competing presuppositions, worldviews, and value systems to yield valid alternative theoretical reconstructions. These believers imply that competing scientific theories are equally able to account for the data and that all that we need to do to obtain the proper scientific theory is to select the right worldview. In other words, the data are perceived as exerting little pressure of their own in moving us toward one or another theory. The data are seen as pliable and equally compatible with different theories. For example, some Christians maintain that the catastrophic global flood theory provides as valid an explanation of geological data as the theory of an old, dynamic Earth. The data can be interpreted as well either way, they say, but Christians should favor the flood theory because it is based on the biblical principle of catastrophism, in contrast to the old-Earth theory, which is based on the unbiblical principle of uniformity.

Other Christians have responded to the new cosmogony more in keeping with the "two books" metaphor than the warfare metaphor, arguing that if modern cosmogony, supposedly based on non-Christian principles, leads to obvious disagreements with the Bible, then clearly any "true" science that is based on biblical principles should lead to harmony with the Word of God. Proponents of the modern "scientific creationism" (creation science) movement claim that they have found that true science. They claim that the biblical framework of a recent creation, a fall into sin with catastrophic physical results, and a catastrophic global flood provides the legitimate framework of interpretation for the data of astronomy, geology, and paleontology.[10] The vast multitude of writings emanating from scientific creationist circles has evidently persuaded a host of lay Christians, teachers, and pastors that the geological and astronomical data really do support assertions that the solar system was created only a few thousands of years ago and that fossils and geological strata were laid down in a global catastrophic flood at the time of Noah. And many Christians believe that "genuine" science—that is, creation science—provides the means for showing the harmony between God's two books.[11]

Scientific creationism has profoundly affected American evangelicalism. Because of efforts to legislate scientific creationism into public school curricula, the issues raised by the movement have even gained the attention of the United States Supreme Court. Scientific creationism had already made strong inroads into the curricula of Christian schools. In addition, many conservative denominations have become permeated with the ideas of scientific creationism. The considerable success of the movement, however, has contributed to additional tensions within evangelical circles. The difficulty is that many Christians are persuaded that scientific creationism is infecting our schools and churches not only with very poor science but with inadequate biblical exegesis as well. Several denominations and many schools, both Christian and public, find themselves entangled in unpleasant controversies over the issue of "origins."

OUR AGENDA

Precisely because of the complex and tense situation within the evangelical community, we humbly offer this volume with the prayer that it may help to point the way toward a resolution of our difficulties. We are

10. See, for example, Henry M. Morris, *Scientific Creationism* (El Cajon, CA: Master Books, 1974). This work presents as good an overview of the basic tenets of the movement as any.

11. In a poll published in 1988, *Christianity Today* reported that of 401 responses from 749 readers of the magazine, 74 percent indicated that they favored the teaching of creationism alongside evolution in public schools. While this high percentage does not indicate that everyone accepts scientific creationism, it does suggest that there is an alarming ignorance of science, of what evolution is, and of the shortcomings of scientific creationism. The poll results were reported in "CT Poll: What Do Christians Want from the Candidates?" *Christianity Today*, 17 June 1988, p. 50.

persuaded that the issues surrounding the biblical doctrine of creation and its relationship to the history, behavior, and structure of the cosmos can be untangled and that progress in understanding and easing the tensions is possible. Because the issues are thorny and varied, we make no pretense of addressing all relevant matters. We deliberately chose to omit discussion of organic evolution and the origin of humankind, fascinating and important as those issues are.[12] Rather we decided to discuss the issues surrounding the inorganic aspect of creation, and we are hopeful that we have laid a solid foundation that may serve as a basis for further discussion concerning the organic realm.

Current controversies over the historically oriented natural sciences and interpretation of the early chapters of Genesis are nothing new. They are part of a very much longer history in which the church of Christ has perennially grappled with the knotty problem of finding satisfactory ways of speaking about the relationships among God, humanity, and the cosmos. Progress in resolving our tensions will not come without considerable understanding of that history. For that reason we offer two historical studies. The first (chap. 2) is a prolegomenon to the study of past efforts to relate Scripture, God, humanity, and the created world. The study illustrates some of the difficulties previously encountered in Christian thinking on these issues. The second (chap. 3) is a specific case study that explores the way in which modern scientific geology emerged from a past in which the study of the Earth was so thoroughly embedded in the concerns of exegesis that progress in understanding Earth history was hindered.

Contrary to the view held by many Christians, we believe that historical reconstructions by modern astronomy and geology are neither uncontrolled speculations nor founded upon unbiblical presuppositions. We hold that these reconstructions are firmly grounded in a wealth of carefully gathered data and have been repeatedly tested by the respected canons of science. To show that the results of these fields are

12. Much as we would like to have dealt with the issues surrounding biological evolution and the origin of humankind, we simply did not have enough time to address these more complex matters. We are persuaded that issues surrounding the inorganic and the organic aspects of creation are closely related, but we caution the reader against drawing unwarranted conclusions. For example, although in this work we adopt the view that the universe is extremely old and has experienced a complex and dynamic history and although we adopt a nonliteralistic approach to Genesis 1, it would be unwarranted to draw firm inferences from these choices concerning our views regarding biological evolution or the historicity of Adam and Eve. Although related, these are distinct issues that must be addressed on their own merits in the light of appropriate biblical, theological, and scientific data.

worthy of the confidence of Christians, we also offer two studies (chaps. 4 and 5) of the methodologies and results of cosmology and astronomy. Furthermore, we have concluded that scientific creationism fails to be "true" science. To that end we have provided a study (chap. 6) of scientific creationism to demonstrate how that approach so violates the canons of sound science that it has become a "sectarian" distortion of science.

Finally and perhaps most important to the majority of our readers, we turn to the issue of what the Bible says about creation. We are convinced that many well-meaning believers operate with inadequate or erroneous views of the biblical doctrine of creation that have needlessly biased them against modern science. Hence we offer an extended essay (chap. 7) that shows how several common perceptions concerning biblical portrayals of God's creative work are not solidly grounded in the text. Our essay focuses on the theological meaning of Genesis 1 and related texts and draws out their powerful implications for the possibility and practice of science.

In conclusion we offer a few pastoral suggestions for applying some of our findings in the context of the pulpit and classroom.

2. HOW DID WE GET HERE?

A Brief Sketch of the Historical Background of the Science-Theology Tension

ROBERT E. SNOW

Christian perspectives on the nature of the created world and on the character of God's activity as the Creator have been shaped not only by Scripture but also by the history of interaction between Christian thought and other intellectual and religious traditions. In this brief historical sketch, Robert E. Snow directs our attention to the influence of both the early Greek concept of Nature and the later professionalization of scientific investigation. Against this background we may gain a clearer view of the contemporary anxiety concerning the relationships among biblical interpretation, creation theology, and natural science.

THE CHRISTIANIZATION OF "NATURE"

To IDENTIFY THE WORLD as God's *Creation* is to recognize the fact that it is completely dependent on the Creator for all things—not only for its inception but also for its governance, purpose, and value. To call the world *nature,* as we often do, is to recognize that it is characterized by an empirically accessible set of ordered properties, behavior patterns, and proximate cause-effect relationships. And when we capitalize *Nature,* we are employing an early Greek, quasireligious concept. Nature (with a capital *N*) is not merely ordered; it is also autonomous—that is, self-ordered. A concept introduced by the Ionian *physikoi* in an attempt to depersonalize the Olympian myths of ancient Greece, Nature itself was invested with some of the attributes normally reserved for deity, such as eternalness, autonomy, and necessity.

The starting point for our study here must be the recognition that the concept of nature cannot be found in either the Old or New Testament. What we encounter again and again as we read Scripture is the claim that the whole world is God's Creation. While the writers of both Testaments exhibit deep interest in exploring the meaning and significance of creation for our understanding of the character of God, our dignity and responsibility as human beings, God's continuing care for his creatures, and the prospects for redemption and new creation (all questions of profound significance to our understanding of our life in Christ), there is a notable lack of systematic discussion concerning the ordered relationships linking phenomenon to phenomenon within the created world. In Scripture, order is presented as a manifestation of God's *covenant* commitment to sustain his Creation.

As soon as Christianity began to move beyond the confines of Judaism and into the Hellenistic culture of the Roman world, the Judeo-Christian faith in the Creator God came into conflict with many contending lines of religious belief. Early evidences of this conflict are given in Paul's minisermons in Acts 14:15-17 and 17:22-31. It is notable that in both instances Paul, in virtually a single breath, presents his God as both the Maker of heaven and earth and the Sustainer of the conditions that make human life possible.

Soon after the close of the apostolic period, one of the issues confronting the patristic theologians was the relationship between their understanding of the biblical concept of Creation and the Greek concept of Nature. During the ensuing polemics, one persistent issue was the tension between various Greek claims about the eternity of the world and the Judeo-Christian understanding of the world as the product of God's creative work. The Christian position increasingly came to emphasize the formula *creatio ex nihilo* and in the process "helped theologians like Tertullian to make the doctrine of creation primarily . . . a question of origins."[1] By the time of Tertullian the theological root of the doctrine of creation had become *creatio ex nihilo,* but the reactions of Tertullian and other Church Fathers to the attempt of Marcion and the Gnostics to separate God the Creator from God the Redeemer helped to

1. Jaroslav Pelikan, "Creation and Causality in the History of Christian Thought," in *Evolution after Darwin,* vol. 3, ed. Sol Tax and Charles Callender (Chicago: University of Chicago Press, 1960), pp. 34-35.

retain the Old Testament emphasis that "the God who acts in history is the Creator."[2]

The major questions raised during the initial stages of the encounter between the concepts of Nature and Creation were more of a metascientific than a scientific character. They were fundamental questions regarding the status of the phenomenal world. Is matter coeternal with the divine? Is the phenomenal world autonomous? Is the order we experience an immanent self-expression of the empirical world, or is it dependent upon God's ordering activity? These questions are essentially metaphysical in character. Their answers have a direct bearing on one's *worldview* and an important, though indirect, effect on the *world pictures* constructed by the sciences.[3] That indirect effect is seen primarily in one's assessment of the legitimacy of scientific investigation and the authenticity of scientifically obtained knowledge.

Just as Nature was a Greek invention, so was metaphysics.[4] Every worldview rests on claims that are metaphysical in character. For example, the claims that a Creator God exists and that the empirical world is completely dependent on him are metaphysical as well as religious claims. Similarly, the Greek claim that Nature is autonomous is a metaphysical claim with important religious implications. Obviously the Greeks were not the first to make metaphysical claims, but they were

2. Pelikan, "Creation and Causality in the History of Christian Thought," p. 34. Pelikan also points out that *creatio ex nihilo* (or its Greek or Hebrew equivalents) may be implied in the Old Testament but it is first found explicitly in the intertestamental period, in 2 Macc. 7:28. In the New Testament it appears in just two passages—Rom. 4:17 and Heb. 11:3.

3. R. Hooykaas has suggested a useful distinction between *worldview* (a set of fundamental beliefs that compose a comprehensive concept of reality—the physical, the spiritual, and their interrelationship) and *world picture* (a set of specific concepts concerned primarily with the contents, structure, and behavior of the physical world). For Hooykaas "the Bible has a certain worldview, that of total dependence of the world on its Creator, but not a definite world picture" (*Religion and the Rise of Modern Science* [Grand Rapids: Eerdmans, 1972], p. 16). Thus the biblical worldview might be consistent with a number of different world pictures (including those embedded in Greek Nature), but it would always be in conflict with a worldview that ascribed to the empirical world attributes such as eternalness and autonomy.

For a more extended discussion of this distinction and its implications for a Christian perspective on scientific investigation, see Howard J. Van Till, "Scientific World Pictures within the Bounds of a Christian World View," *Pro Rege,* March-June 1989, pp. 11-18.

4. *Metaphysics:* The core of metaphysics is reasoned reflection upon what is real, a disciplined concern to identify the basic constituents of reality and the place of humanity in what is real. See William Hasker, *Metaphysics* (Downers Grove, IL: InterVarsity Press, 1983), pp. 13-16.

the first (at least in the Western world) to develop a tradition of making an argued defense and analysis of the metaphysical claims embedded in their worldview. Both science and philosophy, and also much of Western theology, have developed out of the Greek practice of critical assessment of worldview assumptions.

In contrast to the Greek tradition, the biblical tradition asserts rather than argues the existence and action of the Creator God. It was roughly this difference that Augustine apparently had in mind when he wrote,

> When it is asked what we ought to believe in matters of religion, the answer is not to be sought in the exploration of the nature of things, after the manner of those whom the Greeks called "physicists." . . . For the Christian it is enough to believe that the cause of all created things, whether in heaven or on earth, whether visible or invisible, is nothing other than the goodness of the Creator.[5]

The encounter between the Judeo-Christian understanding of Creation and the Greek concept of Nature had momentous consequences. As Christianity became the dominant religion of the West and a major cultural force, Greek worldview assumptions were slowly Christianized.[6] But the Greek tradition also served to shape the development of Christian theological reflection concerning creation. As we noted earlier, for example, Greek claims regarding the coeternity of Nature with the divine encouraged the development of a doctrine of creation in which *creatio ex nihilo* became the central theological focus.

Clearly there are fundamental worldview differences that separate the biblical understanding of Creation from various Greek attempts to develop a coherent account of Nature. Because the radical monotheism of the Judeo-Christian tradition is intimately bound up with the biblical understanding of the world as God's Creation, patristic apologists and theologians soon recognized that they needed to develop some kind of response to the Greek philosophical tradition of attributing a quasidivine status to autonomous Nature. But developing the right response was far from a simple matter. The obvious first step was to deny

5. Augustine, *Enchiridion,* 3.9, Library of Christian Classics, vol. 7, trans. Albert C. Outler (Philadelphia: Westminster Press, 1955), pp. 341-42; quoted by David C. Lindberg and Ronald L. Numbers in "Beyond War and Peace: A Reappraisal of the Encounter between Christianity and Science," *Perspectives on Science and Christian Faith* 39 (1987): 140-49.

6. For further discussion on this, see David C. Lindberg, "Science and the Early Christian Church," *Isis* 74 (1983): 509-30.

that Nature could have attributes that properly belonged to God alone (e.g., eternalness, autonomy, necessity). But to engage in effective argument requires more than simply issuing denials by fiat. To argue that eternalness, autonomy, and necessity should be attributed only to the Creator of the world and not to the principles employed by the Creator to order his Creation is to introduce metaphysics into Christian apologetics and theology. Moreover, since there is but one world to be understood either as Creation or as Nature, and since numerous comments about the world are scattered throughout Scripture, it was clear that patristic thinkers needed to ask whether there was a systematic and consistent biblical alternative to Greek Nature.

Attempts to answer that question required the establishment of some criteria that could be used to determine which biblical statements might be counted as claims about the physical world having direct relevance to the Greek claims. This is essentially a problem involving both epistemology and hermeneutics.[7] We can turn to Scripture for help in developing hermeneutical principles, but the Bible does not provide us with the kind of detailed epistemology that seems to be required. In the absence of a directly revealed epistemology, leading Christian thinkers such as Augustine established by example a pattern that has been followed ever since. Wherever possible, they looked to Scripture, but they also drew upon the thought of both Christians and non-Christians in an effort to understand the problems that confronted them.

Augustine's commentary *The Literal Meaning of Genesis* provides many examples of the impossibility of interpreting Scripture as if it, or the interpreter, existed in cultural isolation. When complaining about the misguided efforts by some Christians to use the book of Genesis as a source of information about the physical particulars of the Creation, he comments,

> It frequently happens that there is some question about the earth, or the sky, or the other elements of this world, the movement, revolutions, or even the size and distance of the stars, the regular eclipses of the sun and the moon, the course of the years and seasons; the nature of the animals, vegetables, and minerals, and other things of the same kind, respecting which one who is not a Christian has knowledge derived from the most certain reasoning or observation. And it is highly deplorable and mis-

7. *Epistemology:* Reasoned reflection on the character of knowledge and how it is to be obtained.

Hermeneutics: The principles of interpretation employed in seeking to comprehend the meaning of the (biblical) text.

> chievous and a thing especially to be guarded against that he should hear a Christian speaking of such matters in accordance with Christian writings and uttering such nonsense that, knowing him to be as wide of the mark as . . . east is from west, the unbeliever can scarcely restrain himself from laughing.[8]

In making this judgment, Augustine clearly believes that he has a reliable epistemological standard by which he can identify authentic knowledge possessed by non-Christians based upon "most certain reasoning or observation" (what we would now probably call "natural science"). Furthermore, Augustine evidently thought that such reasoning or observation could also be used to identify some embarrassing instances of Christians making improper and foolish claims about the teaching of Scripture concerning the created world. Finally, it is evident that Augustine accepted (as we all do in practice) the Greek contention that there does exist a system of relationships that can be used to describe the order of the created world.

Thus the effort to work out the relationships between Nature and Creation brought about far-reaching modifications of the concept of Nature, and it gave a distinctive shape to the Christian doctrine of creation. It also required the development of a Christianized metaphysics and epistemology in order to give appropriate shape to the hermeneutical principles needed to interpret both biblical texts and the Greek claims about Nature. The dynamic interaction of the Judeo-Christian concept of Creation, the Greek concept of Nature, the biblical text, apologetics, theology, metaphysics, epistemology, worldviews, and world pictures that began with the patristic writers has continued to this day.

On the surface it might seem to be a relatively simple task to sort out the relationship between the concepts of Creation and Nature, but in fact it soon became a most complex enterprise that has not yet been completed. Moreover, a strong case could be made that it will never be completed in detail to the satisfaction of all parties involved.

One of the recurring phenomena of intellectual history is the attempt to find a coherent set of relationships linking our understanding of humanity, the divine, and the world. The biblical understanding of Creation and the Greek tradition of reflection concerning Nature overlapped in some ways—they both provided a context for comprehending the phenomenal world of daily experience. Nonetheless, they were parts

8. Augustine, *De genesi ad litteram,* 1.19, trans. Meyrich H. Carre, in *Realists and Nominalists* (London: Oxford University Press, 1946), p. 19; quoted by Lindberg and Numbers in "Beyond War and Peace." For another translation, see our chap. 5, n. 12.

of very different ways of accounting for the relationships among mankind, the divine, and the world. It was not possible merely to add to the biblical concept of Creation the Greek insight that the phenomena of daily experience are linked by an ordered set of relationships that are open to rational exploration. The idea of an "ordered set of relationships" came embedded in metaphysical/religious assumptions from which it had to be untangled before it could be integrated with the Christian understanding of Creation. But the only tools available to accomplish the "untangling" enterprise were those developed by the Greek tradition of metaphysical and epistemological reflection. Although the concern to elaborate a consistent "Christian" view of Nature was far from being a central concern of patristic theologians, they did lay a foundation for recognizing that metaphysics, epistemology, and the search for ordered relationships in "created nature"[9] provided legitimate avenues for Christians to explore—avenues that, as we understand things today, had to be explored in order to understand more fully the meaning of Creation.

Beginning with a series of articles by M. Foster that appeared in *Mind* during the mid-thirties, and represented more recently in a collection of essays entitled *Creation: The Impact of an Idea* and in Eugene Klaaren's book *Religious Origins of Modern Science,* a substantial body of scholarship has developed which argues that the Christian doctrine of creation played an important role in the development of modern science during the seventeenth century.[10] These scholars emphasize the role of medieval voluntarist theology in stimulating a wide-ranging exploration of the consequences of viewing created order as essentially *contingent* (freely chosen by God) rather than *rational* (imposed by a system of logic independent of God). For thinkers within the voluntarist tradition this meant that God's creative activity was not limited in any way by constraints implicit in the Aristotelian (or any other) worldview or world picture. As appropriated by Robert Boyle and others in the seventeenth century, this view of God led to an emphasis on empirical science as the essential tool for learning about "created nature."

By the seventeenth century, the list of Nature's quasidivine at-

9. *Created nature:* A phrase used by the seventeenth-century natural philosopher Robert Boyle to emphasize his understanding of the essential dependence of the natural order on its Creator.

10. *Creation: The Impact of an Idea,* ed. Daniel O'Conner and Francis Oakley (New York: Scribner's, 1969); see especially the essays by Foster ("The Christian Doctrine of Creation and the Rise of Modern Science," pp. 29-53), by Oakley ("Christian Theology and the Newtonian Science: The Rise of the Concept of the Laws of Nature," pp. 54-83), and by Jonas ("Jewish and Christian Elements in the Western Philosophical Tradition," pp. 241-58). Klaaren, *Religious Origins of Modern Science* (Grand Rapids: Eerdmans, 1977).

tributes had been altered: eternalness, autonomy, and necessity were replaced by finitude, dependency, and contingency. The worldview of Nature had been Christianized, and in the process Greek "Nature" and Judeo-Christian "Creation" were merged to provide a major portion of the foundation upon which the structure of modern natural science was to be built. At present there is no scholarly consensus concerning the relative importance of the principal factors responsible for the seventeenth-century scientific revolution. It seems clear that the voluntarist theology of creation played a significant role, but much work remains to be done before its relationship to other significant factors can be fully assessed.

THE PROFESSIONALIZATION OF SCIENCE

By the mid-eighteenth century, science had become so successful a cultural enterprise that its practices were a major resource and standard for metaphysical and epistemological inquiries. To a small but growing number of thinkers it seemed that science could stand on its own feet without the traditional assumption of a Creator God. By the latter third of the eighteenth century there were major examples of both heterodox and orthodox theologies deeply influenced by the growing cultural hegemony of science:

1. The heterodox deists who, in attempting to adjust their understanding of God to Newton's mechanistic universe, concluded that at the beginning God had structured the universe according to immutable physical laws and then left it to function under the governance of those laws. The deists thought it would be contradictory for God to act in history because any action would entail the violation of the laws he had imposed at the world's inception. Consequently they thought it more reasonable and consistent for God to stand apart from his world. A kind of rationality and consistency was thereby gained, but lost was a God personally concerned with each individual and with each part of his Creation.
2. Orthodox Anglicans who, under the leadership of William Paley, elaborated the traditional argument from design into a full-blown theodicy. Starting from the watch-on-the-beach analogy (i.e., the idea that a person discovering a watch on the beach could safely deduce that an intelligence had designed and fashioned it), this

group used the order discovered by science in the natural world to buttress belief in a Creator God. Then they used the assumption of the existence of a benevolent Creator God who intended only good for his Creation to justify existing social, economic, and political arrrangements. Paley's work was extremely influential, but in the context of great stress exerted on the social order by the industrial revolution, and in light of revolutionary changes in natural history concepts as geologists and naturalists incorporated the emerging picture of a dynamic geological environment, it must be viewed as one of the greatest apologetic disasters in the history of Christian thought.[11]

3. Thomas Reid's "common-sense" philosophy, developed squarely on his understanding of Newton's practice and discussion of scientific methodology. Intended as a response to Humean skepticism, Reid emphasized our ability to grasp, without ambiguity, empirical factual information and to proceed inductively to establish true generalizations. His analysis was intended to be normative not only for science but also for historical, theological, and, indeed, all kinds of knowledge. Imported to the United States, it became the foundation for the immensely influential Princeton theology of Hodge and his followers. Hodge thought that his theological method was impeccably scientific and that the dogmatic conclusions it embodied were just as true as the Newtonian science that had served as Reid's model of knowledge. Unfortunately, just as Hodge was putting the finishing touches on his understanding of theological truth, the epistemological foundation of his work was being undermined by William Whewell's development of a radically different hypothetico-deductive model of scientific explanation. The new model was developed in large part to recognize the fundamental role played by unobservable theoretical entities and processes in the best science of the day. While Hodge himself may have had a partial understanding of the extent to which his theological method was dependent on assumptions about the nature of language and about human understanding and scientific method, many of his

11. See James R. Moore, "1859 and All That: Remaking the Story of Evolution-and-Religion," in *Charles Darwin, 1809-1882: A Centennial Commemorative,* ed. Roger G. Chapman and Cleveland T. Duval (Wellington, N.Z.: Nova Pacifica, 1982), pp. 167-94, and "Engines of Empire, Energies of Extinction: Reflections on the Crisis of Faith," paper presented at the Conference on the Victorian Crisis of Faith, Victoria College, Toronto, November 1984.

> followers did not. Unaware of the extent to which their reading of the Bible was conditioned by extrabiblical presuppositions, they often made unwarranted claims for the certainty of their reading of Scripture. This was particularly true of their response to the scientific concept of evolution.[12]

Beginning in the 1830s, Paley's theodicy began to provoke increasing discomfort and even outright hostility among a small but eventually influential group of British intellectuals. The major sticking point was its justification of the status quo as the best of all possible worlds in the face of increasingly widespread distress of both urban and rural populations amid the social trauma of the industrial revolution. One of Darwin's notebooks, composed in the late 1830s shortly after his return to England from the *Beagle* voyage, suggests that a major motive for his attempt to restructure our understanding of natural history was his dismay with the design argument. He wanted to find an alternate account of adaptation that would provide the basis for a more humane theodicy. Because Paley's work was both lucid and tightly reasoned, the web of relationships it spun linking natural history, social policy, and God tended to invite either total acceptance or total rejection. Consequently, when the social setting of English life changed with the unfolding of the industrial revolution, and when the concept of natural history changed under the influence of Lyell and Darwin, it was hard not to reject Paley's God along with his theodicy. The virtually simultaneous decline of the Paleyan design and theodicy perspectives, along with the erosion of the epistemological underpinning of the Princeton theology, left the American evangelical community in substantial disarray.

In the last third of the nineteenth century, Anglo-American science became an autonomous professional activity. In large part it involved a shift in "authority and prestige . . . from one part of the intellectual nation to another."[13] As might be expected, such a shift met with substantial resistance.

12. For an analysis of some of the disastrous consequences of this, see George M. Marsden, "The Collapse of American Evangelical Academia," in *Faith and Rationality,* ed. Alvin Plantinga and Nicholas Wolterstorff (Notre Dame, IN: University of Notre Dame Press, 1983). For an analysis of Reid's role in popularizing "Baconian" science, see "Thomas Reid and the Newtonian Turn of British Methodological Thought," chap. 7 in *Science and Hypothesis,* ed. Larry Laudan (Boston: D. Reidel, 1981). In the same book, see chap. 8 for an important discussion of the issues involved in the development of the hypothetico-deductive method in the late eighteenth and early nineteenth centuries.

13. Frank M. Turner, "The Victorian Conflict between Science and Religion: A Professional Dimension," *Isis* 69 (1978): 359.

> The drive to organize a more professionally oriented scientific community and to define science in a more critical fashion brought the crusading scientists into conflict with two groups of people. The first were supporters of organized religion who wished to maintain a large measure of control over education and to retain religion as the source of moral and social values. The second group was the religiously minded sector of the pre-professional scientific community, which included both clergymen and laymen.[14]

The "professional" party won in part because they were participants in a much broader process of secularization that had been gaining strength in Western society ever since the Renaissance. But their influence was considerably amplified by the inability of those committed to Paleyan modes of argument to distinguish between the timeless and essential elements of their Christian worldview and the obsolete, eighteenth-century static world picture Paley had enshrined in his arguments. Continued attempts to argue for a static world picture because it was *thought* to be consistent with Scripture were increasingly seen as obscurantist and as further evidence that science should become fully "professional." To support their claims for intellectual independence, the professional party created its own "folk" history of science with Darwin and Galileo as its heroes, Bruno as its martyr, and assorted theologians as its villains. The chief and, unfortunately, long-lasting product of this folk history is the "conflict thesis," which employs the militant "warfare metaphor" to describe the relationship between science and religion.[15]

The chief epistemic consequence of the professionalization of science was a change in the status of evidence or ideas drawn from biblical sources. In the mid-eighteenth century it had been perfectly acceptable for Linnaeus to identify his taxonomic category *species* with the *kinds* of Genesis 1. For the great majority of eighteenth-century scientists, such an identification even provided additional scientific stature to the concept. And in early nineteenth-century catastrophic geology it was a legitimate scientific strategy to identify the biblical flood with the last of a series of catastrophes that were thought to account for what was known of the history of the earth. But by the middle third of the century, this was no longer an acceptable argument in the open

14. Turner, "The Victorian Conflict between Science and Religion," p. 364.

15. For a definitive account of the historiographic career of the conflict thesis, see James R. Moore, *The Post-Darwinian Controversies* (Cambridge: Cambridge University Press, 1979), pp. 19-122. For a discussion of the ideological function of folk history and folk science, see Jerome Ravetz, *Scientific Knowledge and Its Social Problems* (New York: Oxford University Press, 1977), pp. 386-97.

scientific literature. Some scientists continued to take account of their understanding of Scripture in their private weighing of the claims of one scientific theory against another, but such "taking account" no longer played a credible epistemic role in the professional publications of geologists. In twentieth-century Western culture, science and its step-brother technology have become the fundamental archetypes of knowledge and understanding. As professional science has broken its ties with the Christian worldview, it has arguably become our most successful and prestigious cultural enterprise. Yet the image of the Christian understanding of the physical universe as God's contingent Creation remains clearly evident in the central role of experiment and observation in modern science, and Christian scientists are still free to follow the example of Robert Boyle and view their work as the disciplined investigation of God's "created nature."

CONTEMPORARY SCIENCE

Within the broader professional scientific community, the accepted rationale for the role of experiment in science is now pragmatic and utilitarian rather than theological. Our culture has learned that experimentation is a fruitful and essential strategy in the pursuit of knowledge concerning nature's properties, behavior, and formative history. Experiment "works," and the manifest success of empirical science is the fundamental justification of experimentation.

But without the constraints of a commonly accepted Christian worldview, the modern idea of nature has a strong tendency to reacquire at least some of the semidivine attributes of Greek Nature. Many people, not all of them scientists, are inclined to attribute autonomy and eternalness to the natural world, and for some, including some well-known popularizers of science, nature functions as an alternative to God in their personal belief system.[16] In many ways we are back to the situation that confronted Augustine. As Christians we have the same challenge of using the available tools of metaphysics to clarify the relationship be-

16. For extended case studies of the way in which popularizers of science often speak as if the universe were self-existent and self-governing, thereby functioning as a substitute for a divine Creator, see "A Masquerade of Science" and "Sagan's Cosmos: Science Education or Religious Theater?" in *Science Held Hostage: What's Wrong with Creation Science AND Evolutionism,* by Howard J. Van Till, Davis A. Young, and Clarence Menninga (Downers Grove, IL.: InterVarsity Press, 1988), pp. 141-68.

tween world picture and worldview and to oppose all worldview claims that conflict with Scripture.[17] We need to sharpen our epistemological and hermeneutical tools in order to assess and compare claims about what "nature" displays and what the Bible teaches. And we need to take to heart Augustine's caution that "it is highly deplorable and mischievous and a thing especially to be guarded against that [an unbeliever] should hear a Christian speaking of [scientific] matters in accordance with Christian writings and uttering such nonsense that, knowing him to be as wide of the mark as . . . east is from west, the unbeliever can scarcely restrain himself from laughing."

There are two major factors that make our task different from Augustine's. The first may seem obvious, but it needs to be emphasized. If we are to speak to our culture as Augustine spoke to his, we must use the functioning metaphysics and epistemology of *our* day rather than those of Augustine or any other early Christian thinker. Perhaps the second point is as obvious as the first, but it too needs emphasis. In Augustine's day the study of nature was only a minor cultural activity, but for us it is the dominant transformer of our culture—so, if the issue was important for Augustine, it is *crucial* for us.

The historical relationships joining Christian faith and the study of the created world are complex and tangled. Obviously, that history has relevance for a number of related intellectual enterprises. Beyond its importance to Western intellectual and social history, what does it tell us about the character of science, how does it bear on our understanding of the doctrine of creation, and what light does it shed on the conflicting claims of contemporary "scientific creationists" and the proponents of evolutionary naturalism? Such questions do not have obvious or simple answers. Even to begin to address them, we need a fitting conceptual framework to guide our reading and analysis. In an important sense, this book is part of a sustained effort to develop and apply a conceptual framework that is capable of enabling us both to appreciate the way in which the Judeo-Christian theology of creation has shaped modern natural science and to critique the numerous and varied claims of twentieth-century proponents of "scientific creationism" and evolutionary naturalism.

17. For a good example of an effort toward clarification of these issues, see Ernan McMullin, "How Should Cosmology Relate to Theology?" in *The Sciences and Theology in the Twentieth Century*, ed. A. R. Peacocke (Notre Dame, IN: University of Notre Dame Press, 1981), pp. 17-56.

3. THE DISCOVERY OF TERRESTRIAL HISTORY

DAVIS A. YOUNG

The static world picture, inherited primarily from early Greek thought and commonly employed in early and medieval Christian writing, was characterized by a basic assumption that the fundamental structures of the world are static. The basic "kinds" of living creatures, along with their terrestrial home, were considered to be relatively fixed. It was assumed that major alterations in these structures came about only through extraordinary and immediate divine action, as in the Noahic flood. And Christians generally believed that the Bible provided the basic outline of the history of those events and phenomena that gave the terrestrial surface its present appearance. However, with the advent of geological science, based on systematic observation and measurement, this concept of Earth history failed altogether to provide an adequate means of accounting for the rich diversity of specific geological phenomena visible to the careful observer. How it failed and what Christians should have learned from this failure are the focus of concern in this chapter by geologist Davis A. Young.

INTRODUCTION

AN IMPORTANT ASPECT of the development of modern science has been the gradual transition from the medieval view of Earth history as brief and rather static to the contemporary view of that history as a long, dynamic succession of events. Over the centuries, discovery after discovery has impressed upon the emerging scientific consciousness a sense of the immensity of the Earth's past. This chapter sketches the story of that growing awareness of the lengthy, complex character of terrestrial history.[1] From this story several themes emerge that are crucial to a satisfactory understanding of the relationship between biblical interpretation and natural science:

1. The decline and disappearance of *concordism*—that is, the assumption within the community of practicing naturalists and geologists that the content of acceptable geological theories must agree with Scripture as understood at the time.[2]

1. As indicated, this chapter constitutes only a sketch of the historical development of geology; for treatments of this subject in greater detail, we recommend the following: Claude C. Albritton, *The Abyss of Time* (San Francisco: Freeman, 1980); Roy Porter, *The Making of Geology* (Cambridge: Cambridge University Press, 1977); Stephen Toulmin and June Goodfield, *The Discovery of Time* (New York: Harper & Row, 1965); Henry Faul and Carol Faul, *It Began with a Stone* (New York: John Wiley, 1983); Francis Haber, *The Age of the World: Moses to Darwin* (Baltimore: The Johns Hopkins Press, 1959); and Martin J. S. Rudwick, *The Meaning of Fossils*, 2nd ed. (New York: Science History Publications, 1976). Two fine brief essays that relate to the substance of this chapter can be found in *God and Nature*, ed. David C. Lindberg and Ronald L. Numbers (Berkeley and Los Angeles: University of California Press, 1986): Martin J. S. Rudwick, "The Shape and Meaning of Earth History," pp. 296-321; and James R. Moore, "Geologists and Interpreters of Genesis in the Nineteenth Century," pp. 322-50.

2. *Concordism* refers to efforts to harmonize the findings of geology, astronomy, and biology with the early chapters of Genesis. The fundamental principle of concordism is that the early chapters of Genesis provide a skeletal outline of historical events that, in principle, can be discovered independently through historical reconstruction by scientific methods. For example, it is assumed that the sequence of events of geological history coincides with the order of events of the six creative days of Genesis 1. A distinction may be made between "literal" or "strict" concordism and "broad" concordism. The former interprets Genesis in a very literalistic manner and expects that scientific data and theories will agree with such an interpretation. The latter recognizes that the data of geology and other historical sciences cannot be reconciled with a literalistic rendering of the text and therefore accepts a variety of figurative, symbolic, or otherwise loose readings of Genesis—such as the idea that the "days" of Genesis 1 may be interpreted as long periods of time.

Support for both kinds of concordism eroded because of their failure to meet essential criteria of modern science outlined by Ernan McMullin in an important article entitled "Values in Science" (*PSA 1982: Proceedings of the 1982 Biennial Meeting of the Philosophy of Science Association*, vol. 2, ed. Peter D. Asquith and Thomas Nickles [East Lansing, MI: Philosophy of Science Association, 1982], pp. 1-25). McMullin enumerates a number of criteria that operate as what he calls "epistemic values" in scientific theory choice, among which are predictive accuracy, internal coherence, external consistency, unifying power, and fertility. Key here is the criterion of external consistency, which McMullin defines as "consistency with other theories and with the general background of expectation." It is assumed, for example, that a good "geological" theory will be externally consistent with other bodies of knowledge. Because in the seventeenth century it was taken for granted that a literal rendering of Genesis yielded valid knowledge of the past, it was assumed that a good "geological" theory would have to be consistent with that external

2. The decline and disappearance of the notion that Scripture supplies useful geological data.
3. The transference of various forms of concordism to Christian apologetic science from a developing professional geology that no longer found concordism relevant to its concerns.[3]
4. The ultimate failure of all forms of concordism as satisfactory Christian solutions to the relationship between biblical exegesis and scientific study.
5. The emergence of the contemporary world picture of professional geology.

THE MEDIEVAL PERIOD

The Islamic world a millennium ago had begun to develop a sense of Earth history. From observation of sedimentation processes and the obvious stratification in many mountainous regions, the Arab thinker Avicenna proposed that the layered rocks of mountains had originated on the sea bed from layers of sediment brought down to the sea by ancient rivers which had in turn derived their suspended sediment load from previously existing landmasses. The sea bed, in Avicenna's view, was ultimately lifted up to form the presently existing mountains. Erosion of these mountains would supply new sediment to be transported by rivers to the sea bed, thus contributing to strata from which future mountains would be formed.[4] Very likely the influence of Aristotle's view of an eternal world on Islamic thought opened up the possibility for thinking about such a dynamic view of terrestrial development.

Avicenna's ideas, however, were neglected for several hundred years. His approach to the Earth was foreign to the medieval Christian mind, for which Scripture, as the infallible revelation of God, was the

---

body of knowledge. As the validity of the knowledge gleaned from a literal reading of Genesis was increasingly questioned, the criterion of external consistency with Scripture slowly fell by the wayside.

3. The term *apologetic science* here refers to science that is employed to lend credence and support to a particular ideology, whether it be a religious position, political view, or social outlook. In the case of concordism, the findings of geology were used apologetically in support and defense of a particular view of the Bible and of Christianity. Conversely, much apologetic science in the nineteenth century could be used to justify the legitimacy of mainstream professional geological conclusions to a conservative Christian audience.

4. For Avicenna's comments on geological topics, see *A Source Book in Medieval Science*, ed. Edward Grant (Cambridge: Harvard University Press, 1974), pp. 616-20.

primary source of knowledge about the history of our terrestrial home. The biblical accounts of creation, the fall, and the flood were generally understood to be literal descriptions of Earth-shaping and Earth-transforming events effected by the immediate activity of God. As miraculous events, the creation, the primeval curse, and the deluge were not described or explained in terms of the efficient cause-and-effect relationships fundamental to modern scientific thought.

Secondary support for the occurrence of these major biblical events was found in the writings of classical historians. Stories of a primeval golden age and of Deucalion's flood were seen as echoes of the events revealed in Scripture. The Earth itself was not regarded as a significant source of knowledge about its own history. Although patristic or medieval writers occasionally referred to the presence of shells in the mountains as an evidence of the deluge, the availability of a divinely inspired history of the Earth rendered any effort to reconstruct terrestrial history from physical evidence contained within the Earth itself unnecessary, fruitless, and unthinkable. Nor did the idea of reconstructing a terrestrial history apart from or preceding human history occur to the medieval Christian mind.

Other factors also prevented the development of terrestrial history reconstructed from physical evidence within the Earth. The medieval worldview was an amalgam of Christian beliefs derived from Scripture and philosophical ideas obtained from the Greeks. The idea of creation was, of course, derived from Scripture, and history was assumed to be only a few thousand years long. The medieval Christian did not doubt that the world had a beginning; any notion of Aristotelian eternalism was rejected along with Aristotle's notion of unending cyclical interchanges of land and sea.[5] Given the brevity of history, it was assumed that changes in life forms or the shape of the land would be negligible. The created order was understood to be essentially static. Integral to this view of a static creation was the notion of plenitude: it was unthinkable that any new life forms could come into existence or that any life forms could ever cease to exist in God's perfect creation.[6]

Whole blocks of Greek thought were combined with the idea of

5. The Greek view of interchanges of land and sea can be found in Aristotle, *Meteorologica* (London: Heinemann, 1952).

6. The notion of plenitude concerns the fullness of God's creation. It was assumed that every being that could exist did exist, that creation was complete. No new beings needed to come into existence, and no beings could pass out of existence, for that would imply a defect or an incompleteness in the creation. If an organism were to become extinct, a link would be removed from the great chain of being.

static creation in the medieval worldview. The cosmos was a hierarchy of entities extending from the perfect heavens down through the terrestrial world of change and decay to hell below. The heavens exerted powerful influences on terrestrial objects.[7] There were, of course, the obvious effects of the Sun and Moon on life, growth, and tides, but the stars were also presumed to exert strong influences. Ristoro d'Arezzo asserted that the heights of mountains were determined by the heights of the stars above them in the sky. Heavenly influences, including the Sun's rays, were presumed to extend into the Earth's interior, affecting the growth of physical objects within rocks as well as vegetation: stony objects that today are readily regarded as the petrified remains of plants or animals were believed to grow in place within rock owing to these influences. That these stony resemblances to living things could grow where they did was as comprehensible as such other examples of growth in stone as hot spring deposits and stalactites in caves.

Because medieval thinking stressed the importance of formal causes, the resemblances and correspondences in form among various objects held great significance.[8] The boundary between organic and inorganic objects was blurred. Human organs were shaped like some stones dug from the Earth. The human body produced gallstones and kidney stones. It was assumed that stones could exert healing influences on the human body. The bubbles and inclusions within crystals were considered analogous to the organs in plants and animals. Geodes—hollow, roughly spherical rocks lined with crystals or banded mineral matter—were thought to be capable of reproduction. A geode containing a loose crystal would rattle when shaken, and when it was cracked open the loose crystal that had grown inside popped out. These geodes were believed to be pregnant female stones about to give birth to baby crystals. Other stones of the appropriate oblong shape were classified as males. Growth, reproduction, organs—stones obviously possessed some degree of life. Why should they not be shaped like other living things?

The organic-inorganic distinction was also blurred with regard to the Earth as a whole. Groundwater movement within the Earth was

7. For a brief discussion of the influences of the heavens on geological features, see Frank D. Adams, *The Birth and Development of the Geological Sciences* (New York: Dover, 1938), pp. 62-68, 78-84.

8. The Greeks had postulated four different kinds of causes: formal, final, material, and efficient. Medieval science was largely built around formal and final causes. Formal causes concerned the forms or ideas that gave shape to objects. Final causes (teleology) concerned the purposes served by objects—that is, the end or goal for which they existed.

On the significance attributed to the similarity of forms, see Adams, *The Birth and Development of the Geological Sciences*, pp. 68-112.

likened to fluid movements within the human body. The ocean could be compared to human sweat. The Earth as a whole could even be considered capable of breathing. This Greek notion was developed by Aristotle, who suggested that the daily movements of vapors into and out of the Earth caused wind and earthquakes in a manner similar to human respiration and movements of gas inside the body.[9]

Medieval thinking was also greatly concerned with teleology. An object was not satisfactorily explained unless its purpose could be ascertained. Stones were often explained in terms of purpose. It was believed that the heavens not only exerted powerful influences on the growth of stones but that they also imparted a variety of occult "virtues" to them. One of the major contributions to medieval understanding of the Earth was the *lapidary,* an exhaustive listing of known stones, including what we would designate as gems, metals, minerals, fossils, concretions, and even pearls. Each stone was described in terms of its virtues and purposes. A given stone might be said to help in childbirth, give success in hunting, prevent blindness, prevent intoxication, aid in prayer life, or yield clues to the faithfulness of one's spouse.

In view of this explanatory framework, it is not surprising that the medieval world had no conception of terrestrial history. The universe was a God-created static world order of interlacing affinities of form and purpose. In the absence of any strong sense of efficient cause-and-effect relationships, there was no interest in developing a notion of historical sequence recorded in the rocks. Since fossils were thought to grow in place in rock, their use as historical documents of past life forms was not recognized.[10] Moreover, with the grand events of terrestrial history described in Scripture, there was no impetus to work out that history empirically from evidence in the rocks.

THE RENAISSANCE

As the medieval synthesis began to crumble during the Renaissance, students of the natural world became increasingly dissatisfied with explanation in terms of formal and final causes. Observation, measure-

9. Aristotle, *Meteorologica.* See also Adams, *The Birth and Development of the Geological Sciences,* pp. 426-45.

10. The term *fossil* originally meant anything that was dug out of the ground. Through the years the term has become restricted largely to the remains of once-living organisms that are embedded in rock or unconsolidated sediments.

ment, description, and explanation in terms of material and efficient causes received greater emphasis. Agricola, for example, emphasized careful description of the physical properties of stones and downplayed the significance of their virtues.[11] Besides, close observation of some stones indicated that the assignment of their virtues had been groundless. This demagicalization of stones opened up a greater tendency to treat them as objects to be measured, weighed, and described.

Stones that bore close resemblance to living creatures received greater attention. Scholars such as da Vinci, Fracastoro, Fallopio, Colonna, Aldrovandus, Gesner, and Palissy all pondered the character of these "formed stones."[12] While differing about the origin of the stones, they invoked material causes to explain their presence, suggesting that they were the remains of organisms destroyed by the flood, that they were the remains of organisms that had left the sea and been stranded on land, or that they were the result of organic seeds that had been placed in rock fissures by sea spray and subsequently grown there. Even those who regarded fossils as inorganic were more likely to attribute their origin to material fluids than to heavenly influences.

The invention of printing and the development of methods for publishing book illustrations brought about an explosion of information about stones.[13] Illustrations removed the need for applying unreliable imagination to already obscure verbal descriptions. Readers could see what the authors were discussing and could begin to appreciate the crystalline shape of true minerals or the remarkably fine and detailed

11. See Georgius Agricola, *De Natura Fossilium,* trans. M. C. Bandy and J. A. Bandy (New York: Geological Society of America, 1955). The original Latin work was published in 1546 and represented the most thorough compilation of knowledge about various kinds of stones. It broke with the older traditions of mineralogy by concentrating on the physical characteristics of minerals rather than on their supposed virtues.

12. "Formed stones" was a commonplace designation for what today we call fossils. For an excellent discussion of how scholars of the Renaissance period dealt with "formed stones," see Rudwick, *The Meaning of Fossils,* pp. 1-48. Rudwick points out that much of the problem regarding the origin of "formed stones" was really a problem of determining which stones should be classified as having an organic origin and which might be classified as having formed *in situ* through the agency of fluids acting on rock. As Rudwick points out, many stones that bear a resemblance to plants and animals are not of organic origin. Moreover, the concept of extinction was not yet accepted during the period in which these early naturalists worked, and although many of the "formed stones" looked very much like genuinely organic materials, they nevertheless had no known counterpart in the modern world.

13. Rudwick discusses the importance of illustration in advancing understanding of the nature of fossils in *The Meaning of Fossils.* Woodcut illustrations, introduced in the mid-sixteenth century, represented a significant improvement over purely verbal description, but the copper engravings that began to appear in the early seventeenth century allowed for representation of a far greater wealth of detail.

resemblances of "formed stones" to known living creatures. Students of the Earth also began to engage in systematic collection of stones. Specimens had formerly been collected as random grab samples, but now they were carefully described in terms of their local settings. These technical advances and improvements in communication and collection led to improved accuracy in description of terrestrial objects.

The revolution in astronomy also changed the character of Earth study. Galileo's discoveries about the Moon, sunspots, and the satellites of Jupiter strongly suggested that the Earth was part of a coherent Sun-orbiting planetary system. The new astronomy also undermined notions of the Earth's spatial uniqueness and led to speculation not only that the Earth might be just one body among several in a planetary system but also that the solar system might be just one of many planetary systems orbiting a multitude of suns. In *Principles of Philosophy* (1644), René Descartes proposed a scenario for the development of Earth and Earth-like bodies from cooling suns in thoroughly mechanistic terms.[14] In doing so, he opened the way for subsequent attempts to explain the origin of the globe mechanistically. Even seventeenth-century discussions of the origin of the globe that were motivated by dedication to the absolute authority of Scripture began to combine biblical data with mechanistic descriptions and natural law.[15]

Much of the research into the Earth during the sixteenth and seventeenth centuries concerned the description of various classes of stones, but a few naturalists turned their attention to the Earth's history. While such great naturalists as Martin Lister, Agricola, Edward Lhwyd, John Ray, and Conrad Gesner expended most of their "geological" energies on the nature of stones, Robert Hooke and Nicolaus Steno began to explore historical problems.

Steno was particularly interested in the relationship between organic-looking stones and the character of their host rocks and in the relationships of stratified rocks to one another. He eventually worked out some basic principles of rock stratification and attempted to reconstruct the local "geological" history of Tuscany in Italy on the basis of

14. Descartes, *Principles of Philosophy*, trans. V. R. Miller and R. P. Miller (Boston: Kluwer, 1983). Part IV of the *Principles* is devoted to a discussion of how the Earth and similar bodies could develop by means of the interaction of various kinds of particles. Descartes also discusses the nature of the atmosphere, oceans, mountains, earthquakes, and tides.

15. A prime example, discussed below, is that of Thomas Burnet, who tries to explain the deluge in terms of second causes. He flatly rejects any notion of a miraculous creation of waters for the flood. Moreover, he constructs an original Earth in terms of the coagulation of particles in much the same way that Descartes had.

the relative positions and characteristics of strata.[16] Although Steno generally attributed the stratified rocks to the agency of the flood, he developed his historical reconstructions and principles of stratification in orderly, mechanical, causal terms. He reasoned from the effects seen in the rocks to their causes by analogy with present causes that would produce similar effects. He suggested, for example, that a stratum containing "marine productions" would have formed in the sea. He did not attribute the features he observed to occult influences or the immediate supernatural activity of God.

Robert Hooke, like Steno, argued at great length for the organic origin of organic-looking "formed stones." In addition, Hooke wanted to explain causally just why such remains are found where they are, in rocks far removed from our present oceans. Like Steno, he declined to attribute these features to divine action, though neither did he deny the possibility of such an origin; rather, using a mechanical approach, he contended that successive earthquakes had gradually elevated former sea beds to their present mountainous positions.[17]

BRITISH DILUVIALISM

In seventeenth-century Great Britain the study of natural history developed rapidly. Although scholars were already communicating the fruits of their labors with one another through letters, the founding of the Royal Society of London and the publication of its *Philosophical Transactions,* beginning in 1665, promoted a stronger sense of a scholarly community devoted to natural history. Increased opportunity for travel and field study also spurred the growth of natural history. Much of the scholarly work, exemplified by the studies of such individuals as Martin Lister, John Beaumont, Edward Lhwyd, Thomas Molyneux, and Abraham de la Pryme, focused on the nature of "formed stones" or unusual rock formations such as the Giant's Causeway in Ireland.[18]

16. See Steno, *De Solido intra Solidum Naturaliter Contento Dissertationis Prodromus* (Florence, 1669). An English translation of Steno's great work was reprinted by Hafner Press in 1968. Of great importance in Steno's *Forerunner* was the inclusion of the first schematic cross-sectional diagrams designed to illustrate the geological history of a region.

17. See Hooke, *Lectures and Discourses of Earthquakes, and Subterraneous Eruptions* (1705; reprint ed., New York: Arno Press, 1978).

18. The first hundred years of the *Philosophical Transactions* contained many articles pondering the nature of fossil remains. Regarding the Giant's Causeway, see S. Foley, "An Account of the Giants Causway in the North of Ireland" 18 (1694): 170-82; Thomas

These naturalists generally avoided theorizing about the Earth and its history, working instead to accumulate detailed knowledge about the smaller objects.

British theorizing about Earth history or large-scale terrestrial features generally developed within the context of natural theology. The Earth and its features were interpreted in light of the conviction that the Earth is a grand theater of divine benevolence, wisdom, contrivance, and purpose. This interest in combining natural history and natural theology spawned a host of "physico-theological" treatises in the late seventeenth and early eighteenth centuries.[19] These works treated the history of the globe and many of its features (e.g., metals, mountains, "formed stones," and waters) within the biblical scheme of creation, fall, flood, and final consummation. Most of these physico-theological discourses stressed the importance of a global cataclysmic flood in shaping topographical features, in forming strata, and in accounting for fossils—an approach to terrestrial history I will subsequently be referring to as *diluvialism*.

Diluvialism was unquestionably the dominant scheme of terrestrial history in seventeenth- and early eighteenth-century Britain, and its effects were felt even on the continent, as in work of Scheuchzer and Lehmann.[20] *Within the diluvialist school of late seventeenth- and early eighteenth-century Britain, it was taken for granted that a valid theory of the Earth must be consistent with a generally literalistic reading of the creation and flood narratives in Genesis. Since it was assumed that Scripture supplied important reliable data that could be used in construction of a theory of terrestrial history, it was thought that there must be an external coherence of theories of the Earth with biblical data.* Other useful sources of information included the writ-

---

Molyneux, "Some Additional Observations on the Giants Causway in Ireland," 20 (1698): 209-22; and Richard Pococke, "An Account of the Giants Causeway in Ireland," 45 (1748): 124-27, all in *Philosophical Transactions*.

19. Such treatises typically integrated various natural phenomena with a full-blown natural theology. Representative of the genre was John Ray's *Three Physico-Theological Discourses* (London: Innys, 1713).

20. J. J. Scheuchzer was a German naturalist who argued compellingly in his *Fishes Complaint and Vindication* (Zurich, 1708) that fossil fish skeletons were not mere sports of nature but remains of animals that had been trapped in sediment during the flood. Indeed, Scheuchzer felt that virtually the entire sedimentary rock pile was diluvial. He was so enamored of diluvialism that he spent much of his career searching for the remains of humans who had been buried by the flood. Skeletal remains were eventually found that Scheuchzer attributed to the *Homo diluvii testis*, the human witness of the flood, but several years after Scheuchzer's death, the paleontologist Cuvier showed that the remains were those of an extinct giant salamander! Johann Lehmann made valuable contributions to the advancement of stratigraphy through his labors in detailed mapping of successions of strata in Germany, but nonetheless held that the layered sedimentary rocks were deposited in the Noahic deluge.

ings of the classical historians and artifacts dug from the Earth. The British naturalists of this period assumed that there was a *literal concordism* between actual Earth history and the history of the Earth as narrated in the Bible.

The immediate impetus for development of the diluvialist outlook was the publication of Thomas Burnet's *Sacred Theory of the Earth* in 1681. Burnet set out to write a history of the globe that would trace out a "system of natural providence" and explain the causes that produced the world. The account was biblically based because, Burnet wrote, "the Sacred writings of Scripture . . . are the best monuments of Antiquity."[21] Following Descartes, Burnet proposed that the Earth had developed from a primordial chaos. The original particles of matter coagulated into a sphere whose outer surface was perfectly smooth and without mountains or sea. The original Earth had a concentric structure in which the solid crust rested on a subterranean sphere of water. Ample biblical demonstration of the reality of this subterranean abyss of waters was provided by such texts as Psalms 24:2 ("The earth was founded upon the seas, and established upon the floods"), 136:6, and 33:7. The Sun's heat, penetrating into the interior of the Earth, led to gradual expansion and cracking of the underside of the crust. At the time of the Noahic deluge, the waters of the abyss poured out through the fractured crust onto the surface of the Earth. Burnet maintained that the subterranean abyss was the only satisfactory solution to the perennial problem of finding a source of water sufficiently large to cover the entire globe. Preferring to formulate a solution solely in terms of second causes, Burnet dismissed the idea of a purely miraculous creation and subsequent annihilation of waters just for the flood. In the absence of any other rational means to account for the necessary quantity of water, he was left with the theoretical need for the abyss. The fractured crust split into several large fragments which foundered and tilted in the subterranean waters, he suggested, and this tilting created the mountains. He associated the smoothness of the globe at the time of its creation with the perfection of original creation, and he saw the mountains as evidence of disorder on the Earth's surface, evidence of the ultimate disordering catastrophe, the flood.

While Burnet's structure of terrestrial history received its main architectural lines from Scripture, it was buttressed by the ancient historians and minor topographical and physical evidence. In some instances Burnet felt the need to give new interpretations to biblical texts

21. Burnet, *The Sacred Theory of the Earth* (1681; reprint ed., London: Centaur Press, 1965), p. 24.

in order to fit the demands of the theory. His work triggered a host of rebuttals and similar hypotheses. The result of these responses was the establishment of a long tradition of diluvialist cosmogonies that provided an integrative framework for many of the discoveries accumulating from the study of the Earth.

Some of Burnet's critics claimed that he was subverting true religion by adopting new interpretations of the biblical text. One of the most vigorous responses to Burnet was the *Geologia* (1690) of Erasmus Warren. Warren piled exegesis upon exegesis to show that a host of the theory's details flew in the face of biblical truth. Yet even Warren, totally convinced as he was of the historical veracity of the biblical account of the creation and the flood, could not resist trying to explain the cause of the flood. Opposing the ideas of the abyss and the original perfect smoothness of the globe, Warren theorized that the floodwaters had been held in large caverns within mountains. But to save his own rational flood theory, Warren also fell prey to latitudinarian interpretation. He felt that it was a big mistake to assume that the entire surface of the globe was submerged. Rather than insist that the highest mountain peaks were submerged by at least fifteen cubits, Warren suggested that only the general land surface had been covered to that depth and that mountain peaks could very well have risen above the mantle of water.

After the publication of Newton's *Principia* in 1687, a number of authors sought to work out a global cosmogony within the diluvialist framework that would also incorporate the new physics. The two leading efforts along this line were those of John Woodward and William Whiston. Woodward of Cambridge was a first-rate naturalist who had intimate field acquaintance with both fossil remains and rock stratification. Through his own travels and correspondence with other European naturalists, Woodward was persuaded that stratification was a global characteristic. Consequently, his diluvialism stressed the importance of rock stratification as well as fossils. In *An Essay towards a Natural History of the Globe* (1695), Woodward suggested that the entire outer portion of the Earth was "dissolved" into its constituent particles at the time of the flood. Although he envisioned the abyss as the source of the floodwaters, he did not propose a specific mechanism for its migration to the Earth's surface. For a time the surface was totally covered by a slurry of water, earthy particles, and organic remains, he said. At the conclusion of the flood, the suspended particles settled gravitationally according to their densities. These particles were deposited into strata. Dead organisms likewise settled in order of specific gravity, which accounted for fossils in stratified rocks. Although Woodward claimed that his hypothesis was built on observation and induction, it is clear that his commitment to biblical and natural theology controlled the direction in which his inductions led. Beyond maintaining that the flood was a historical event demanded by the biblical record, he also asserted that the accompanying "destruction of the Earth was not only an Act of the profoundest Wisdom and Forecast, but the most monumental proof that could ever possibly have been, of the Goodness, Compassion, and Tenderness, in the Author of our Being," since the flood rendered the Earth suitable to the condition of fallen mankind.[22]

William Whiston, following Newton and Edmund Halley, suggested in *A New Theory of the Earth* (1696) that the close approach of a comet tilted the Earth on its axis at the time of Adam's fall and that a massive downpour of water from the tail of a later comet was responsible for the flood. Whiston followed Woodward in stating that fossils and stratification were the result of sedimentation of materials at the conclusion of the flood.

Diluvialism received one of its most capable treatments in 1761 from Alexander Catcott in *A Treatise on the Deluge.*[23] Catcott held that the deluge plainly left its marks on nature and that nature clearly showed the handiwork and wisdom of its Creator. He provided a thoroughly scriptural backing for his theory, but, like his predecessors, he was obliged to give peculiar twists to certain biblical texts so that they would mesh with an otherwise thoroughgoing causal explanation of the deluge—as, for example, in his interpretation of the source of the flood waters. With Burnet and Woodward, Catcott believed that a subterranean abyss was the source of the flood, and he appealed to many of the same texts in support of that contention. However, Catcott's explanation of the derivation of the abyss, revealed in his interpretation of the second day of creation, was peculiarly unorthodox. He maintained that God fashioned the firmament as a dual structure, consisting of external and internal parts. The external firmament corresponded essentially to the atmosphere, and a corresponding internal firmament was in the interior of the Earth at the original creation. On the second day of creation, he wrote, God separated the waters of the oceans from the air, making them the waters under the firmament. But at the same time, God also separated waters from the internal firmament, creating a watery

22. Woodward, *An Essay towards a Natural History of the Earth* (1695; reprint ed., Arno Press, 1978), p. 94.

23. For a very informative and penetrating study of the contributions of Catcott, see Michael Neve and Roy Porter, "Alexander Catcott: Glory and Geology," *British Journal for the History of Science* 10 (1977): 37-60.

subterranean abyss suspended above an airy internal firmament, making them the waters above the firmament. Catcott's published cross-section of the Earth showed air above, then ocean, then a solid rocky crust, then a subterranean watery abyss, then air, and finally the central fire. This theory led to the curious result that the waters above the firmament were physically beneath the waters below the firmament.

At the time of the flood, wrote Catcott, the waters of the subterranean abyss were forced through fractures in the crust by the expansion of the internal firmament. He maintained that the opening of the floodgates of heaven had nothing to do with torrential downpours of rain, but rather referred to the floods of water being expelled from below by the internal "heavens" or "airs." These waters flowed over the world destroying life, dissolving the Earth's surface in Woodwardian fashion, and then redepositing great uniform thicknesses of essentially horizontal, layered, fossiliferous sediments. Mountain peaks and intervening valleys were carved out of the evenly layered sediment cover of the globe as the floodwaters drained off the surface back into ocean basins and ultimately to the abyss.

Catcott's treatise, representative of the diluvial genre, appealed not only to Scripture but also to available physical evidence and to such historical traditions as the flood legends and the Atlantis story. Data were obtained from any source deemed reliable. In Catcott's case the physical evidence included not only the Earth's stratification and the fossil remains to which Woodward had appealed but also widespread deposits of surficial gravels, erratic boulders, bone-filled cave deposits, and the shape of such topographic features as river valleys, which in England are commonly much wider than the streams flowing through them.

EIGHTEENTH-CENTURY DESCRIPTIVE GEOLOGY

In continental Europe, naturalists were generally less constrained by natural theology in their view of the world than were the British. The grand-scale cosmogonies of the continent generally were not cast within a framework of biblical history as were those of their British counterparts.[24] Continental students of the Earth tended to be more concerned

24. An example of the grand-scale cosmogony is that of Comte de Buffon. Buffon's ideas were published in a series of volumes issued through the mid-eighteenth century entitled *Histoire Naturelle*. His final contribution to theorizing about the Earth was *Epoques*

with careful description and classification of strata and mountains than with speculation about terrestrial history. The pervasiveness and importance of rock stratification, dimly perceived in Great Britain by Woodward, de la Pryme, Bellers, Strachey, Catcott, and Michell, were brilliantly illuminated by geological labors in Italy, France, and Germany throughout much of the century.[25] Nevertheless, authors of these theories often attempted to show at least a broad agreement between the Bible and geological theory in order to make the theory more palatable to a potentially skeptical audience.

During the early eighteenth century, J. J. Scheuchzer and Antonio Vallisnieri described and sketched local successions of complexly distorted alpine strata.[26] Anton Lazzaro Moro, impressed by the formation of islands in the Mediterranean Sea by recent volcanic eruptions, concluded that all islands and mountains were ultimately volcanic in nature.[27] Moro recognized that the higher peaks in the heart of many mountain ranges are typically composed of unstratified rocks such as granite. He referred to these as "primitive" mountains. He also recognized that the lower ranges and foothills are commonly composed of stratified rocks that are superimposed on the primitive materials; these he termed "secondary" mountains.

---

de la Nature. Buffon postulated that the Earth had cooled from material torn from the Sun by a passing comet. On the basis of several cooling experiments on heated spheres of various materials, Buffon calculated that it would require at least 132,000 years for the Earth to cool off to its present condition. To make peace with theologians, Buffon suggested that there had been seven major epochs in the history of the globe and also that the "days" of Genesis 1 should be regarded as periods of time of indeterminate length.

25. For examples of the British contributions (in addition to Woodward's *Essay towards a Natural History of the Earth* and Catcott's *Treatise on the Deluge*), see Abraham de la Pryme, "Concerning Broughton in Lincolnshire, with his Observations on the Shell-Fish, observed in the Quarries about That Place," *Philosophical Transactions of the Royal Society of London* 22 (1700): 677-87; Fettiplace Bellers, "A Description of the Several Strata of Earth, Stone, Coal, etc. Found in a Coal-Pit at the West End of Dudley in Staffordshire," *Philosophical Transactions of the Royal Society of London* 27 (1710): 541-44; John Strachey, "A Curious Description of the Strata Observ'd in the Coal-Mines of Mendip in Somersetshire," *Philosophical Transactions of the Royal Society of London* 30 (1719): 968-73, and "An Account of the Strata in Coal-Mines, etc.," *Philosophical Transactions of the Royal Society of London* 33 (1725): 395-98; and John Michell, "Conjectures Concerning the Cause and Observations upon the Phenomena of Earthquakes," *Philosophical Transactions of the Royal Society of London* 51 (1760): 566-634.

26. The diluvialist J. J. Scheuchzer drew some diagrams of the contorted strata of the Alps that Antonio Vallisnieri reproduced in his *Lezioni Accademica Intorno all'origine delle Fontane* (Venice, 1715) and discussed further in his *De Corpi Marini che su Monti si Trovano* (Venice, 1721).

27. A brief discussion of Moro's ideas can be found in Adams, *The Birth and Development of the Geological Sciences*, pp. 365-72. The original source is Anton Lazzaro Moro, *De Crostacei e Degli Altri Marini Corpi che si Truovano su Monti* (Venice, 1740).

Johann Gottlob Lehmann concluded that mountains could be classified into three categories.[28] He, too, spoke of primitive mountains, composed of crystalline rocks rich in vein materials, and attributed them to the original creation. Superimposed on the primitive mountains were the "stratified" rocks *(Flötzgebirge)*, essentially equivalent to the secondary mountains of Moro. In the German district of Thuringia, Lehmann traced a succession of thirty superimposed stratified formations belonging to the *Flötzgebirge* over a large distance. His publications included careful diagrammatic cross-sections of the Thuringian stratigraphic sequence.[29] Lehmann suggested that the Noahic flood might have produced these strata by eroding materials from the already existent primitive mountains. Lehmann's third category included lesser mountains formed by volcanic action and smaller-scale floods.

In 1759, Giovanni Arduino, working independently of Lehmann, came to similar conclusions about primitive and secondary mountains.[30] He further stressed that the strata of the secondary mountains are commonly tilted and fossiliferous. Arduino's third category of "tertiary" materials included generally horizontal unconsolidated sediments (e.g., sand and gravel) that underlay low mountains and hills. He also included volcanic lavas in this group. And he proposed a fourth division of rocks that included surficial alluvial materials derived by erosion from the mountains and carried downhill by streams. The German naturalist Pierre Simon Pallas also recognized a threefold division in the Ural Mountains.[31]

Much valuable stratigraphic description was carried out during the eighteenth century. The simple three-part classification of primitive, secondary, and tertiary mountains and rocks found increasing acceptance as a means for description of stratigraphic successions and for correlation of widely separated areas. By the late eighteenth century it was widely accepted that primitive rocks consisted mainly of crystalline granite, schist, and gneiss, devoid of fossils but rich in veins and metal

28. See Lehmann, *Versuch einer Geschichte von Flötzgeburgen* (Berlin, 1756).

29. "Stratigraphic sequence" refers to the physical succession of rock layers that are superimposed one on top of the other much as a set of books or papers could be piled in a stack. The discussion of a stratigraphic sequence would include information such as the rock type(s) composing the layers, their thickness, and possibly their weathering characteristics.

30. Arduino's ideas were published in 1759 in *Osservazione sulla fisica constituzione delle Alpi Venete* and in two letters to Vallisnieri printed under the title *Saggio Fisico-Mineralogico di Lythogomia e Orognosia*.

31. See Pallas, *Observations sur la Formation des Montagnes et les Changements arrives au Globe, particulierement de l'Empire Russe* (St. Petersburg, 1777).

deposits, that occurred in the heart of mountainous country and that cropped out on the highest peaks. Superimposed on top of these highly complex, contorted rocks were thick successions of secondary rocks, distinctly stratified and generally composed of fossiliferous material such as sandstone, limestone, coal, chert, and shale. The secondary strata typically cropped out in the foothills of the higher mountains and were commonly tilted with respect to the horizon. The tertiary strata, superimposed more or less horizontally above the beveled edges of secondary strata, were generally found in the valleys at the base of mountains and were fossiliferous and commonly unconsolidated.

Coincident with the development of stratigraphic classification were many detailed local studies that demonstrated that stratified rock piles, measured perpendicular to the layering, were in some places thousands of meters thick! Moreover, it was recognized that within these sedimentary piles the order of succession of strata was unvarying over wide regions. Individual strata, even very thin ones, were found to be laterally persistent over considerable distances, and some of the strata gave evidence of having been deposited under tranquil conditions.[32]

THE COLLAPSE OF DILUVIALISM

The discovery of the great thicknesses and regularity of strata cast serious doubt on the diluvialist framework. Naturalists had increasing difficulty in understanding how even a year-long flood could possibly erode enough material to deposit thousands of meters of sediment and how a turbulent flood could account for the persistent regularity of very thin strata.

Another strain for diluvialism came from studies of volcanism in central France. The naturalist Jean Etienne Guettard visited the Auvergne region of central France in mid-eighteenth century. Impressed by the black building stone used in the villages, Guettard was directed by the local citizenry to quarries from which the rock had been excavated. He was able to trace the rock from the quarries to large cone-shaped landforms that resembled modern-day volcanoes in every respect. The summit craters, however, were now completely covered with vegeta-

32. For an extensive compilation of many of the eighteenth-century studies of local geology, consult the classic work on the history of geology by Karl A. von Zittel, *Geschichte der Geologie und Palaeontologie* (Munich, 1899).

tion, and some had sheep grazing in them. Guettard reasoned that if the cones were volcanic, they obviously had been inactive for a long time.[33] The Romans who once occupied the area left no written records about volcanic eruptions, nor were there any legends of such activity in the area. Hence these volcanoes must be remnants of terrestrial activity that was quite ancient in terms of the generally accepted time-frame for the Earth. This discovery led to a vague sense that the Earth might have had some history of its own prior to human occupation. *Human history and Earth history were no longer coterminous. Earth history needed to be separated from human history.*

The implications of Guettard's discovery were further heightened by the field work of Nicholas Desmarest in the same area in the 1760s.[34] His more detailed studies of the volcanic cones and lava flows showed that there had actually been a succession of eruptions. He was able to show from analysis of the distribution of landforms and lava flows that the most recent flows had followed the courses of previously existing river valleys and were later subjected to renewed downcutting by the streams that he found flowing across them. Moreover, the lava-filled channels had been carved into yet older lavas that could be seen capping present-day terraces at higher elevations than the lava-filled valleys. The sense of a terrestrial antiquity incompatible with diluvialism was heightened, and the divorce between human history and a prehuman geological history was sharpened.

Discoveries about stratigraphy and volcanism were not the only source of tension for diluvialism. Indeed, the original fruitfulness of diluvialism contained the seeds of its downfall. The biblical flood story had provided the basis for research into the history and structure of the Earth that stimulated both biblical interpretation and empirical study of the Earth. For example, diluvialism promoted acceptance of the organic origin of fossils and encouraged naturalists to examine more closely the details of stratification, the distribution of organisms within strata, and the character of landforms such as valleys and mountains. But both biblical and geological studies introduced tensions that helped to undermine diluvialism.

On the one hand, geological field work stimulated by the diluvial

33. The classic paper on the discovery of volcanic activity in central France is Guettard's "Memoire sur quelques Montagnes de la France qui ont ete des Volcans," *Memoires de l'Academie Royale des Sciences* (1752): 27-59.

34. On this, see Desmarest, "Memoire sur l'Origine et la Nature du Basalte a Grandes Colonnes Polygones, determinees par l'Histoire Naturelle de cette Pierre, Observee en Auvergne," *Memoires de l'Academie Royale des Sciences* (1771): 705-75.

hypothesis led to discoveries that could not be explained satisfactorily in terms of the flood. For example, it was easy enough to test Woodward's hypothesis that the flood deposited sedimentary materials in order of decreasing specific gravity. Almost as soon as the ink was dry on Woodward's *Essay,* naturalists such as Hauksbee showed that strata are not arranged in any particular order of specific gravity.[35] Abraham de la Pryme was puzzled by field observations that marine and terrestrial fossils were not contained in the same layers as they should be if a global flood had occurred.

Alexander Catcott's unpublished notebooks also disclose the severe problems for diluvialism posed by field observation.[36] According to Catcott's theory, the flood deposited essentially horizontal layers of sedimentary rock in Woodwardian fashion. The problem was that he frequently encountered exposures in which the strata were severely tilted and contorted. How could the flood do that? Even more serious was the existence of conglomerate layers. It was thought that such layers had clearly been deposited by the flood, and yet they contained pebbles and boulders of sandstones and limestones—rock types that were themselves supposed to be the product of flood action. It almost seemed to demand two flood stages separated by a dry period: one flood to assemble layers of sand and lime mud, which upon drying would have produced sandstone and limestone, and another flood to deposit pebbles of these rock types into the conglomerate layers. Catcott had also proposed that valleys had been excavated as the floodwaters drained off the face of the Earth. The difficulty was that many valleys had been excavated out of extremely hard, consolidated rock. While a flood could excavate soft sedimentary rocks, how could it erode granite? A host of such basic observations led to increasing dissatisfaction with diluvialism as a way to account for the Earth's features.

Severe tensions were also being introduced by a growing number of competing exegetical interpretations of relevant biblical texts. The mechanism of the global flood had been variously identified with a subterranean abyss (Burnet, Woodward, Catcott), overflow of the oceans onto sinking continents (Hooke), caverns in mountains (Warren), tidal pull of the ocean onto land by a passing comet (Halley), drenching rains from the tail of a comet (Whiston), and displacement of the oceans by a

35. Fr. Hauksbee, "A Table of Specifick Gravity," appended to an article by F. Bellers, "A Description of the Several Strata of Earth, Stone, Coal, etc. Found in a Coal-Pit at the West End of Dudley in Staffordshire."

36. See Neve and Porter, "Alexander Catcott: Glory and Geology."

realignment of the center of the Earth (Ray). Interpretation of the events of the second day of creation had led to such diverse proposals as Catcott's two firmaments and various identifications of the firmament with the sky, the crust of the Earth, or the interior of the Earth. This wide diversity of interpretations raised serious doubts about the usefulness of Scripture as a source of detailed historical information about the flood or the prediluvian state of the world.[37]

By the late eighteenth century, diluvialism had rightly disappeared from the knowledgeable geological community. Naturalists were beginning to realize that Scripture was not a particularly useful source of data about the history of the deluge. And because of the plethora of conflicting "literal" interpretations of the flood story, it was no longer considered necessary for a good geological theory to be consistent with a generally literal understanding of the creation and flood narratives. *Literal concordism clearly had not worked.* However, because of the Christian commitment of a significant portion of working naturalists and geologists, future geological theorizing still paid close attention to the possibility of some kind of concord between geology and Scripture. Interpretations of biblical texts were broadened in order to achieve agreement with the results of empirically based geological investigations. In the late eighteenth and early nineteenth centuries there was still a sense among many geologists that a valid theory of the Earth must be consistent with some interpretation of the creation narrative and that the flood narrative in Genesis must at least refer to a local inundation. It was still assumed, despite the demise of diluvialism, that there must be a *broad concordism* between actual Earth history and the history of the Earth recorded in the Bible.

Since the late eighteenth century, observational evidence against diluvialism has continued to accumulate, and that model of Earth history has consistently been found wanting. As long ago as 1834 the great Christian geologist and ordained minister Adam Sedgwick charged the authors of the "Mosaic Geology" of his day with having committed "the folly and the sin of dogmatizing on matters they have not personally examined, and, at the utmost, know only at second hand—of pretending to teach mankind on points where they themselves are uninstructed."[38]

37. For a more thorough discussion of the various diluvialist reconstructions of earth history, see Davis A. Young, "Scripture in the Hands of Geologists (Part One)," *Westminster Theological Journal* 49 (1987): 1-34.

38. Sedgwick, quoted by Edward Hitchcock in "The Connection between Geology and the Mosaic History of the Creation," *The Biblical Repository and Quarterly Observer* 6 (1835): 266.

And a year later, Christian geologist and theologian Edward Hitchcock wrote that diluvialism "has been abandoned by all practical geologists."[39] That observation remains in force today. Yet surprisingly diluvialism has survived to the present in the form of scientific creationism.[40] But there are good reasons why scientific creationism falls outside the geological enterprise and why the active geological community continues to reject diluvialism. Two and a half centuries of vigorous research into God's world have uncovered a plethora of data that are totally incompatible with the notion that the world's stratified rocks were deposited by a single deluge.

NEPTUNISM

Some Renaissance naturalists such as da Vinci and Fracastoro had been skeptical that the flood could adequately account for the distribution of fossil remains. Such doubts continued into the seventeenth and eighteenth centuries, so that diluvialism was not without theoretical rivals. Aware of the classical Greek speculations of Herodotos, Aristotle, and Strabo about interchanges of land and sea, some scholars of the seventeenth and eighteenth centuries tentatively suggested similar hypotheses to account for fossils. Still others developed a scheme, ultimately termed *neptunism,* that attributed fossils and strata to progressive diminution of a world ocean. For a time this scheme successfully replaced diluvialism and provided a new productive research program.

In *Protogaea* (1749), the philosopher Leibniz proposed that the Earth had cooled through time from a molten state, that during the cooling process a universal ocean had condensed from vapor, and that this gradually condensing and cooling ocean left deposits in the form of fossil-bearing rocks.[41]

39. Hitchcock, "The Connection between Geology and the Mosaic History of the Creation," p. 284.

40. Scientific creationism is a modern form of diluvialism that seeks to interpret all the data of geology in terms of a literalistic reading of Genesis 1 (seven twenty-four-hour days in succession) and of the flood narrative (a global cataclysm that killed virtually all life on earth). The movement is typified by such institutions as the Institute for Creation Research and the Creation Research Society and by such individuals as Henry M. Morris and Duane Gish.

41. G. W. Leibniz, *Protogaea* (Göttingen, 1749). A sketch of Leibniz's ideas of earth history had been available as early as John Woodward's *Essay towards a Natural History of the Earth,* but the full manuscript was published posthumously.

In the early eighteenth century, the French ambassador to Egypt Benoit de Maillet, strongly influenced by his contact with Arab scholars, speculated about the deposition of fossiliferous rocks during gradual diminution of the sea. He was aware that his theories were radical, and, anticipating a hostile reaction, he presented them in the form of a fantasy entitled *Telliamed,* attributing the ideas to the character of an Indian philosopher.[42] His thesis had a fairly strong empirical base and sought to interpret features within layered fossiliferous rocks by analogy with modern-day sedimentation processes on ocean bottoms. Moreover, he supported his theory with clever experiments and measurements of that rate at which the sea might fall. In all, the theory was creative and viable, but it drastically broke with the common assumptions about the age of the globe: the original manuscript suggested that the planet might be as old as two billion years! De Maillet's friend and publisher reduced this unimaginable number to two million years in the published version, but even this smaller figure constituted an affront to Christians accustomed to the idea of a six thousand-year-old Earth. *Telliamed* was also plainly at odds with other more central Christian ideas, and in spite of an attempt to reconcile its conclusions with a somewhat liberalized interpretation of Genesis, the book had little lasting impact on the community of natural philosophers.

Later in the century, Georges Louis Leclerc, Comte de Buffon, an ardent Newtonian, set out to do for our understanding of the globe what Newton had done for our understanding of the solar system. After performing several cooling experiments on heated spheres composed of different materials, Buffon developed an elaborate scheme of Earth history along the lines of Leibniz and de Maillet in which the Earth gradually cooled to the point where it became habitable. Like some of his predecessors on the continent, Buffon believed that the fossiliferous rocks had been formed by sedimentation from a gradually receding ocean. Buffon's theory proposed an Earth several tens of thousands of years old, and, like de Maillet, he sought to reconcile his views with the Bible by adopting a metaphorical interpretation of the days of creation.

Neptunism received refined treatment in the hands of Abraham Gottlob Werner and his disciples.[43] Blessed with rare teaching ability, a

42. Benoit de Maillet, *Telliamed; or, Conversations between an Indian Philosopher and a French Missionary on the Diminution of the Sea,* trans. Albert V. Carozzi (Urbana: University of Illinois Press, 1968). The first edition of *Telliamed* was printed posthumously and anonymously in 1748. De Maillet conceived the title for his book by spelling his own name backwards.

43. Werner has commonly been regarded by geologists as obstinately biased in his views, a perception that can be traced to Charles Lyell's influential treatment of the history

systematic mind, a position in the Mining Academy at Freiberg, Saxony, and numerous gifted students, Werner occupied the first academic position devoted entirely to "geognostical" learning.[44] This set of circumstances helped make Wernerian neptunism geological orthodoxy throughout continental Europe, Great Britain, and even North America until the early nineteenth century. Building on the labors of Arduino, Lehmann, and others, Werner stressed the importance of stratigraphic sequence to "geognosy." His stratigraphic classification was essentially the same as that of Lehmann and Arduino, but he added two new classes to the widely accepted threefold classification. A fourth category of "transition" rocks *(Übergangsgebirge)* was sandwiched between the primitive and the secondary (or, following Lehmann, *Flötzgebirge*), and a fifth class of rocks included recent surficial and volcanic materials.

Werner theorized that the primitive crystalline rocks of the high mountains had been formed by chemical precipitation from a primeval global ocean rich in dissolved chemicals. He believed that the veins in the mountains had likewise been formed in fissures by chemical precipitates derived from the ocean above. As the sea diminished, he suggested, portions of the primitive rocks were exposed above sea level and were partly eroded, providing mechanically derived detritus.[45] The transition rocks were stratified materials deposited on the flanks of the primitive mountains by a combination of chemical precipitation and mechanical sedimentation. Only a few organic remains were found in the transition rocks. As the land masses increased owing to diminution of the sea, mechanical erosion increased. Hence the secondary *(Flötz)* layers were largely clastic sedimentary deposits containing abundant remains of a variety of life forms.[46] Werner's tertiary deposits *(Aufge-*

---

of geology. In fact, Werner was one of the outstanding geological minds of the eighteenth century. For more sympathetic treatments, see Alexander M. Ospovat, "The Distortion of Werner in Lyell's *Principles of Geology*," *British Journal for the History of Science* 9 (1976): 190-98; and Mott T. Greene, *Geology in the Nineteenth Century* (Ithaca: Cornell University Press, 1982), pp. 19-45.

44. Werner preferred to use the term *geognosy* to *geology*. To Werner, geognosy connoted genuine knowledge of the earth, whereas geology entailed too great a sense of the speculative and hypothetical. Werner and his disciples commonly referred to themselves as geognosts rather than geologists.

45. The term *detritus* is used in geology to denote fragments of rock material that have been worn from larger fragments or outcrops. Gravel, pebbles, sand, and clay are all different kinds of rock detritus that have been derived in part by the mechanical weathering of other materials. Mechanical weathering is contrasted with chemical weathering and implies formation of the fragments by breaking, grinding, splitting, and so on rather than by chemical reaction.

46. The term *clastic* means "broken." Clastic rocks are those that are composed

schwemmte Gebirge) were generally unconsolidated sediments containing the remains of advanced life forms. Werner regarded volcanism as a very recent phenomenon; volcanic rocks were essentially superimposed on top of the tertiary deposits. He also believed that the rock type basalt, now known to be a volcanic rock, was a chemical precipitate from the diminishing ocean. Hence basalts in primitive, transition, or secondary deposits could not be used as evidence for ancient volcanism. Werner felt that his stratigraphic system, although admitting of some local exceptions, was generally valid worldwide.

Wernerian neptunism successfully filled the late eighteenth century void left by a dying diluvialism. Neptunism was imported into Great Britain, where it was integrated into the demands of natural theology and "scriptural geology" in spite of the fact that Werner never mentioned God or the Bible in his writings. Although the total stratigraphic column was no longer attributed to the Noahic flood as in diluvialism, some geologists interpreted the upper part of the column (unconsolidated, vertebrate-bearing surficial gravels and sands and erratic boulders spread across northern Europe) as monuments to the universal deluge. In addition, with its assumption of a primeval ocean and emerging continental landmasses, neptunism was readily adaptable to the creation account of Genesis 1. Late eighteenth-century Christian naturalists from the British Isles could be just as dogmatic that the neptunist scheme was the friend of orthodoxy as the seventeenth-century diluvialists had been regarding the capabilities of the Noahic flood.

Richard Kirwan, for example, a staunchly orthodox Irish Protestant naturalist, vehemently defended neptunism against other systems he deemed atheistic. "Geology," he wrote, "naturally ripens, or . . . *graduates* into religion, as this does into morality."[47] In his view, geology meant neptunism, and neptunism was the scheme taught in Scripture. Genesis 1:2 was a description of the primeval universal ocean that progressively receded until dry land eventually appeared on the third day of creation. Moses' accuracy as a historian was established because he taught about Werner's ocean. The Spirit of God moving upon the waters of the great deep was an "invisible elastic fluid" rather than the third person of the Trinity. More specifically, the darkness on the face of the deep in Genesis 1:2 alluded to a "great evaporation that took place

primarily of broken fragments of previously existing rocks and minerals. *Clastic* and *detrital* are roughly synonymous.

47. Kirwan, "On the Primitive State of the Globe and Its Subsequent Catastrophe," *Transactions of the Royal Irish Academy* 6 (1797): 234.

soon after the creation as soon as the solids began to crystallize."[48] Chemical precipitation and release of heat were implicit in the biblical account. Psalm 104:6, speaking of waters standing above the mountains, referred to the primeval ocean depositing crystalline solids contained in the chaotic waters to form the primitive mountains. Later came the flood, a great wave washing over the northern continents from a source to the south. For Kirwan, as for the diluvialists, Scripture still provided the broad framework for Earth history, but the details of that framework had changed considerably from diluvialism in order to meet the demands of the new theoretical outlook on Earth history as provided by geological investigation. Biblical interpretation, even among the most rigidly orthodox, continued to react and adjust to extrabiblical concerns. The result was a broadening concordism.

DEVELOPMENTS OF THE EARLY NINETEENTH CENTURY

As the nineteenth century began, geology became a distinct intellectual discipline in response to the explosion in knowledge about the Earth. As the century progressed, the development of this science was aided by the establishment of geological societies and publications, government-sponsored geological surveys, and academic chairs devoted strictly to geology. Geology was becoming increasingly professional, and a consensus was forming that the proper way to decipher Earth history was to reconstruct it from the physical effects of past events preserved in the rocks. Those events could be discerned by assuming that processes that produce known geological effects today would have produced the same effects in the past. Far less attention was paid to Scripture than had been the case a century earlier.

Very late eighteenth- and early nineteenth-century studies contributed to the demise of neptunism, further demonstrated the erroneous character of the old diluvialism, and led to a developing appreciation for the magnitude of geological time. In 1795 James Hutton argued that the "primitive" rocks were not relics of the original creation but that they too had a history.[49] He showed that many of the layered

48. Kirwan, "On the Primitive State of the Globe and Its Subsequent Catastrophe," p. 266.

49. Hutton's landmark ideas were published in his *Theory of the Earth* (Edinburgh: W. Creech, 1795). What he called primitive rocks are today called metamorphic rocks. Locally he found instances in which layers of metamorphic schists were interbedded with

primitive rocks might simply be altered sedimentary rocks originally deposited from water and later buried and heated. Moreover, he demonstrated that many so-called primitive granites did not underlie the secondary rocks, as neptunism taught, but had been intruded into them in a molten condition and were therefore younger than the secondary rocks. Though many of Hutton's points fell on deaf ears at the time, his views were more successfully advanced and developed later by John Playfair's *Illustrations of the Huttonian Theory* (1802) and Charles Lyell's *Principles of Geology* (1830-33).

Among the most important advances of the late eighteenth and early nineteenth century was the linking of paleontology and stratigraphy. Earlier naturalists such as Lister and de la Pryme had noted that there was a connection between types of fossil remains and the kinds of rocks that enclosed them. Yet it remained for William Smith, Georges Cuvier, and Alexandre Brongniart to demonstrate the close relationship between stratigraphic units and their contained fossil species. In the 1790s, William Smith, a canal engineer, mapped in detail the Secondary strata of most of England.[50] He carefully worked out the succession of superposed strata and was able to trace individual formations for great distances across the English countryside. Moreover, he noted that each stratigraphic unit was characterized by specific fossil remains. He published his findings in a table of formations and a geological map of England in 1815.[51] Though Smith paid little heed to theoretical explanations of the stratigraphic succession he had mapped, most geologists, impressed by the order, thicknesses, extent, and regularity of fossil content, interpreted these rocks as deposits resulting from repeated interchanges of land and sea.

---

layers of less altered rocks that nonetheless contained fossil remains, and from that he concluded that the metamorphic rocks were altered sediments.

50. The terms *secondary* and *tertiary* were used with such regularity that they eventually became formal terms in a developing geological time scale. As such they are generally capitalized (and will be subsequently in this chapter). The term *Tertiary* is still used in the geological time scale, although *secondary* and *primitive* have vanished from use.

51. Smith's map was formally entitled "A Geological Map of England and Wales, with part of Scotland; exhibiting the Collieries, Mines, and Canals, the Marshes and Fen Lands originally overflowed by the Sea, and the varieties of Soil, according to the variations of the Substrata, illustrated by the most descriptive Names of Places, and of Local Districts; showing also the Rivers, Sites of Parks, and principal Seats of the Nobility and Gentry, and the opposite Coast of France." It was published on fifteen large sheets, covering a total area of 8 feet 9 inches by 6 feet 2 inches. The scale was 5 miles to an inch. He also published "A Geological Table of British Organized Fossils, which identify the Courses and Continuity of the Strata" in 1815. The map and table went through a number of subsequent modifications and editions.

In an 1808 publication, the French comparative anatomist Georges Cuvier, in collaboration with geologist Alexandre Brongniart, accomplished for Tertiary deposits what Smith did for the Secondary. Especially interested in vertebrate anatomy, Cuvier studied the vertebrate-rich Tertiary sedimentary rocks of the Paris basin, which were known to overlie the bulk of British strata. He recovered remains of several vertebrates of hitherto unknown species and established the principles of vertebrate skeletal reconstruction. Cuvier was finally able to demonstrate that extinction was a reality in the world, once and for all putting to rest the medieval doctrine of plenitude.[52]

Cuvier and Brongniart further demonstrated that the order of succession of sediments in the Paris basin remained the same over great distances. Although locally some units might be missing from the sequence, the relative position of strata was always the same, and most units, including many thin layers, were traceable for long distances. Moreover, Cuvier and Brongniart showed that limestone layers containing marine invertebrates were interlayered with quadruped-bearing continental deposits. Simple neptunism was inadequate to explain such alternations, which quite clearly pointed to interchanges of land and sea. They also found that the quadruped remains varied from one continental deposit to another, suggesting that there had been not just one, but several episodes of extinction. Cuvier proposed that the Earth had experienced a number of great revolutions in which the sea alternately invaded the land, wiping out animal populations, and retreated, allowing the land to be repopulated by animals from other parts of the globe.[53]

52. It is true that there remained some resistance to the idea of the extinction of species. While many invertebrates present in the fossil record were still unknown in the modern world, one could always argue justifiably that such life forms might yet live undiscovered in the deep recesses of the sea. And in fact species were occasionally found that were known first from the fossil record. But Cuvier stressed that it was unlikely that large land quadrupeds could have been overlooked during the substantial exploration of the Earth's continents that had taken place by the early nineteenth century. Even if a couple of species had been missed, it was unlikely that a large number of species had been overlooked—especially some of the dramatic forms of which Cuvier had reconstructed the skeletons: the mastodon, the giant sloth, and a giant salamander. These discoveries presented yet another difficulty for the global flood theory: if Noah had in fact preserved each animal species by taking at least two of each kind on board the ark, how was one to account for the evidence of so many extinct species?

53. See Cuvier, *Essay on the Theory of the Earth* (Edinburgh: Blackwood, 1817). The volume cited here is the third edition of an English translation of Cuvier's work. The original French work formed the preliminary discourse of Cuvier's monumental four-volume work (published in 1812) on the bones of fossil quadrupeds.

Subsequent work in Europe by other geologists disclosed further evidence of physical breaks at levels in the strata that also contained major changes in fauna. Consequently the idea that terrestrial history had been interrupted by several revolutions or catastrophes became widespread. In this brand of catastrophism, however, the relatively brief catastrophes were believed to have punctuated otherwise lengthy periods of calm deposition.

Continental catastrophism, like the earlier diluvialism and neptunism, was adapted to the concerns of British natural theology. Concerned over the loosening of ties between geology and Christianity, William Buckland, an Oxford geologist and Anglican minister, sought a new harmonization between Scripture and the physical world in catastrophism.[54] Although committed to an old Earth, Buckland believed in catastrophes and specifically identified the most recent catastrophe with the Noahic flood. He forcefully argued, as had Catcott and Kirwan, that erratic boulders and widespread surficial gravel deposits were the result of the flood. Moreover, he insisted that the river valleys of England were too large to have been excavated by their own streams and so must have been scoured by the action of the flood. Buckland's primary evidence came from recent discoveries of numerous fissure and cave fillings that contained abundant quadruped remains; he argued that these too had been washed to their final resting place by the sweeping waves of the flood.[55] Many other geologists also attributed *surficial* deposits to a nearly global or continent-sized flood.

In the meantime, field research at Auvergne was turning up evidence damaging to both neptunism and catastrophism. Wernerian neptunism had stressed that volcanic activity was strictly recent and that basaltic rocks were not volcanic but chemical precipitates. The preliminary work of Guettard, Desmarest, and others in Auvergne had already produced evidence that volcanic activity was ancient and that basalt was a volcanic rock, but the result of their work was not yet widely appreciated. New studies in the early nineteenth century by Daubeny, Scrope, and Lyell and by such Wernerians as von Buch and d'Aubuisson de Voissons established conclusively that basalt was *not* an oceanic chemical precipitate but rather lava.[56]

54. For a stimulating analysis of the establishment of geology at Oxford by William Buckland, see Nicolaas A. Rupke, *The Great Chain of History: William Buckland and the English School of Geology, 1814-1849* (Oxford: Clarendon Press, 1983).

55. Buckland, *Reliquiae Diluvianae* (London: John Murray, 1823).

56. Charles Daubeny published several papers on the volcanoes of Auvergne. Among these were "On the Volcanoes of the Auvergne," *Edinburgh New Philosophical*

These studies also further intensified the growing sense of the Earth's antiquity. Like his predecessors, Daubeny recognized different periods of volcanic activity at Auvergne, and yet, still working with one eye on the biblical flood, he distinguished between ancient (prediluvian) and modern (postdiluvian) eruptions and erosion. Scrope's more detailed work in the early 1820s, later reinforced by that of Charles Lyell, stressed that the widely accepted deluge that supposedly swept over Europe and carved out river valleys upon its retreat could not have affected the Auvergne region. Scrope argued, as Lyell did later, that Daubeny's older "prediluvian" volcanic cones, composed of very loose cinders, would have been obliterated by the effects of such a deluge. Moreover, Scrope argued that the lava-filled valleys had been excavated by the action of flowing streams rather than a flood. It was Scrope's contention that the phenomena pointed to a continuum of eruptions and stream erosion over a long period of time. There was no justification from the field evidence for any clear-cut distinction between prediluvial and postdiluvial times. Moreover, the physical remains of the succession of volcanic eruptions were superimposed on top of a succession of thinly layered Tertiary beds containing fossil remains of a number of still-living invertebrate species. And the Tertiary beds, when traced out of the region, could further be seen to be superimposed above the very thick accumulations of earlier Tertiary and Secondary rocks. Scrope sensed that the amount of time necessary to account for the features he was observing had to be vastly greater than what naturalists had previously accepted.[57]

Despite Buckland's efforts to resurrect the idea of a great flood, even English naturalists were becoming restive in the 1820s over efforts

---

Journal 3 (1820): 359-67 and 4 (1821): 89-97, 300-315, and "On the Diluvial Theory, and on the Origin of the Valleys of Auvergne," *Edinburgh New Philosophical Journal* 10 (1831): 201-29. Also of great importance is George Poulett Scrope's *Memoir on the Geology of Central France, Including the Volcanic Formations of Auvergne, the Velay and the Vivarais* (London, 1827), a considerably altered second edition of which was issued in 1858 under the title *The Geology and Extinct Volcanoes of Central France.* Lyell's work in the Auvergne also presupposes the volcanic origin of the basalts. See Charles Lyell and Roderick I. Murchison, "On the Excavation of Valleys, as Illustrated by the Volcanic Rocks of Central France," *Edinburgh New Philosophical Journal* 12 (1829): 15-48. For an analysis of some of the early nineteenth-century work on the volcanoes of Auvergne, see Martin J. S. Rudwick, "Poulett Scrope on the Volcanoes of Auvergne: Lyellian Time and Political Economy," *British Journal for the History of Science* 7 (1974): 205-42.

57. Scrope's famous memoir on volcanoes climaxes near the end with the sentence, "The leading idea which is present in all our researches, and which accompanies every fresh observation, the sound which to the ear of the student of Nature seems continually echoed from every part of her works, is—Time!—Time!—Time!"

to force geology to submit to the demands made by diverse interpreters of the Bible. Those efforts always foundered on the rocks! Geologists wanted geology to stand on its own merits and come to its own conclusions on the basis of empirical evidence interpreted in the light of known causes. More than anyone, Charles Lyell urged upon his contemporaries the need to reconstruct terrestrial history solely in terms of geological evidence without regard to biblical data. He was largely successful in showing that many geological phenomena need not be interpreted in terms of catastrophes with religious overtones.[58]

Eventually even Buckland abandoned talk of a global deluge. By appealing to the effects of active alpine glaciers, Louis Agassiz, a Swiss Christian, provided persuasive evidence that the gravels, erratic boulders, grooved and polished outcrops, and wide valleys all over Europe were far better accounted for in terms of the action of widespread glacial ice sheets that had covered large portions of northern Europe than as deposits of a great flood.[59] Upon comparing alpine glacial phenomena with European "flood" deposits, both Lyell and Buckland recognized the superiority of the glacial hypothesis. Global floods were dead in the geological community.

BY 1840, THE GEOLOGICAL COMMUNITY had been led down several dead-end paths by biblically based theories of Earth history, particularly those that called for global floods or global oceans. From its historical experience of a succession of failures of "biblical geology," the professional

58. Charles Lyell's views have commonly been misunderstood. Because he was a classic early advocate of uniformitarianism over against catastrophism, he is often viewed as having opposed the idea that catastrophes could have been involved in any geological processes. While it is true that Lyell believed that many ancient geological features could be explained without recourse to large-scale catastrophes—particularly those having religious overtones (because he feared the introduction of biblical data into the science of geology)—nevertheless it is plain that he was open to the occurrence of rather large catastrophes in Earth history. For example, he talked plainly of the catastrophic outflow of Lake Erie once the Niagara River eroded back to the mouth of the lake.

Lyell's conception of uniformitarianism (a term that he himself did not use) has also been confused. There are several strands to his thinking about geological uniformitarianism, and the reader who wishes to understand what he said ought to read not only his *Principles of Geology* but also Martin J. S. Rudwick, "Uniformity and Progression: Reflections on the Structure of Geological Theory in the Age of Lyell," in *Perspectives in the History of Science and Technology*, ed. D. H. D. Roller (Norman, OK: University of Oklahoma, 1971), pp. 209-27; and Stephen Jay Gould, "Toward the Vindication of Punctuational Change," in *Catastrophes and Earth History* (Princeton: Princeton University Press, 1984).

59. Agassiz, *Etudes sur les Glaciers* (Solothurn, Switzerland: Jent & Gassmann, 1841). Buckland was evidently persuaded of the glacial hypothesis sometime in the 1830s.

geological community, composed of a very sizable proportion of Christians, finally realized that *in the future the science of geology would be hindered by continued adherence to the epistemic value that valid geological theories must be consistent with a literalistic reading of the creation and flood narratives in Genesis.* External coherence of geological theory with biblical data was no longer fruitful for geology as a science. Concordism, whether literal or broad, no longer had any significant place in geology as a science or in the technical geological literature.

The entire practicing geological community was fully convinced that terrestrial history was long, complex, dynamic, and quite probably measurable in terms of millions of years. There was no known geological evidence to controvert that conclusion. The simple stratigraphic classification of the eighteenth century had been subdivided and refined to the extent that a workable geological time scale was being developed.[60] The various stratigraphic subdivisions each represented lengthy blocks of time. Steady-state uniformitarians (Lyell), directionalists (de la Beche), actualistic gradualists (Scrope), and catastrophists (Elie de Beaumont) were all agreed on Earth's antiquity.[61] Geologists might disagree about details of terrestrial history and about mechanisms of mountain build-

60. For a brief overview of the development of the geological time scale, see W. B. N. Berry, *Growth of a Prehistoric Time Scale* (San Francisco: Freeman, 1968).

61. A steady-state uniformitarian (e.g., Hutton and Lyell) believes that there has been no net increase or decrease in the energy or intensity of various geological causes throughout time, that geological history displays a steady state and lacks any directionality or progressive features. Uniformitarians would deny, for example, that volcanic activity was any more intense in the past on average than it is at present or that there has been progression among life forms.

A directionalist believes that there has been a net increase or decrease in the intensity of various geological processes acting through time. A modern directionalist would certainly insist that the meteorite-bombardment process that has affected many bodies within the solar system was far more intense in the beginning than it is now.

An actualistic gradualist believes that all geological processes that have occurred in the past are at least potentially explicable in terms of the kinds of processes that are observable somewhere today, that there no need to appeal to unknowable events to explain various geological phenomena. A gradualist, however, insists that geological processes have always been relatively mild in intensity and that catastrophes, while they may occur, were not more intense in the past than at present. A gradualist could also be a uniformitarian (like Lyell) or a directionalist (like Scrope). An actualist could also be a catastrophist.

A catastrophist (e.g., Elie de Beaumont and Adam Sedgwick) believes that many geological features are not explicable in terms of the kinds of processes presently observable occurring at the rate at which they now occur. Earlier catastrophists often talked of gigantic floods or of cataclysmic mountain uplift; modern catastrophists talk of bombardment of the planet by a large asteroid at the end of the Cretaceous Period. A catastrophist could be a directionalist but never a uniformitarian. A catastrophist could also be an actualist.

For more on these categories, see Rudwick, "Uniformity and Progression," and Gould, "Toward the Vindication of Punctuational Change."

ing, but there was general agreement about the broad sweep of Earth history. With the major exception of Lyell until the latter part of his life, there was also general agreement that life history as reflected in the fossil record was progressive. Even the many pre-Darwinian geologists who opposed the idea of biological transformism (evolution) accepted the progressive character of life on Earth.[62] And by 1840, virtually all geologists agreed that the available evidence no longer supported the notion of a gigantic flood.

Not only did geologists agree about the overall outline of Earth history; they agreed fundamentally about method. By 1840 there was a consensus that geology as a science should reconstruct Earth history solely in terms of geological evidence interpreted by analogy or identity with known causes and effects, without regard to biblical data. Those individuals who promoted "scriptural geology" in the nineteenth century were now outside of the geologic community, left behind by newer discoveries of the science.

Clearly the period from 1680 to 1840 was marked by a drastic change in attitude regarding the relationship between Scripture and scientific study of the Earth. At the outset of the period Scripture was regarded as the primary source of information about the main events of Earth's history. Constantly forced to adjust biblical details to new empirical discoveries from the Earth, Christian naturalists eventually adjusted the character of the framework. Ultimately they came to the conclusion, in the light of extrabiblical data, that Scripture provided neither a major framework of events of Earth history nor even details about events in Earth history. Every effort over the previous two centuries to make Scripture, literalistically interpreted, and geological data conform to one another to the satisfaction of the geological community eventually failed.

62. For example, a widely held idea during the first half of the nineteenth century had been vigorously supported by the French geologist Leonce Elie de Beaumont. He had suggested that large mountain chains such as the Andes had arisen from the sea very abruptly and with such force that the surface of the sea was cast into gigantic waves. For a detailed discussion of such views see Greene, *Geology in the Nineteenth Century*, pp. 69-121.

One might consider here such professional geologists as Adam Sedgwick and Edward Hitchcock and the fine amateur geologist Hugh Miller. In a presidential address to the Royal Physical Society entitled "Geological Evidences in Favor of Revealed Religion," Miller spoke of two extreme schools of his time: "that school which, founding on a certain progressive rise, in the course of the geologic periods, from lower to higher types, both animal and vegetable, would infer that what we term creation is in reality but development . . . and another school . . . which teaches that there has been no upward progress in creation" ("The Old Red Sandstone [New York: Hurst, 1858], pp. 303-14).

SINCE 1840

While the professional geological *community* of the mid-nineteenth century no longer looked for outlines or details of events of terrestrial history in Scripture and was no longer concerned about trying to corroborate Scripture with geological discovery, there were many outstanding *individual* Christian geologists, particularly in Great Britain and North America, who were still interested in the question of how Genesis and the discoveries of geology were interrelated. Their solutions were published in church magazines, theological journals, and popular books but not in the geology journals. *Broad concordism had left the arena of geology as a discipline and entered the arena of "apologetic geology."*

Without exception, these Christian geologists assumed, as had their predecessors, that the creation and flood stories were chronicles of early history. Moses was still regarded as the sacred historian. Genesis 1 was presumed to describe factual events that occurred in a historical sequence. The major problem, of course, was how to reconcile the days of Genesis 1 with the geological knowledge of Earth's extreme antiquity. One approach was the gap theory put forward by Scottish preacher and amateur naturalist Thomas Chalmers and subsequently adopted by geologists William Buckland and Adam Sedgwick in England and Edward Hitchcock in America. This interpretation held that the condition of a perfect creation recorded in Genesis 1:1 lasted for an indeterminate period of time that correlated with the geological ages. An unmentioned time gap intervened between verses 1 and 2, and verse 2 presented a fallen, chaotic Earth that had experienced a catastrophe or ruin, perhaps linked with the fall of Satan. The six days of creation were the literal re-creation, reconstruction, or restitution of the Earth from its fallen, chaotic state to a newly ordered, habitable condition.

A more successful concordistic scheme was the day-age theory, popularized in Scotland by Hugh Miller and Robert Jameson. This theory was adopted by the Scottish theologians James McCosh and James Orr, by North American geologists Arnold Guyot (Princeton), Benjamin Silliman and James D. Dana (Yale), and J. William Dawson (McGill), and by the Princeton thinkers Charles Hodge, A. A. Hodge, and B. B. Warfield.[63] The main assertion of this view was that since there is no biblical record of an evening and morning for the seventh day, the

63. For a somewhat more extensive list of theologians who were very favorable to some form of the day-age theory, see Davis A. Young, *Christianity and the Age of the Earth* (Grand Rapids: Zondervan, 1982), pp. 55-67.

seventh day has not ended yet. It is therefore a long period of time of indeterminate length; so, too, must the other six days of creation be long periods of indeterminate length. It matters not therefore how old the Earth may be, for the Bible really says nothing of its age. Another leading feature of the day-age view was that the succession of events as disclosed by geological investigation was said to be remarkably similar to that outlined in Genesis 1. Day-age concordists never tired of stressing the near identity of the two sequences and viewed that as proof that the Bible is adequately corroborated by geological science. Some concordists, such as Miller and geologist G. F. Wright (Oberlin), also tried to salvage a local deluge by appealing to geological evidence in central Asia or the Middle East.

At present, however, only a relatively small number of Christians continue to work at constructing a detailed correspondence between the Bible and geological history. Their works functions not as normal empirical science but as "apologetic geology." Their writings are published almost entirely in church publications, theological journals, or popular literature but not in the geological journals. The gap theory has been abandoned on virtually all fronts. There has been a remarkable resurgence of literal concordism within the scientific creationism movement, but not among Christian professional geologists. A few Christian geologists have favored the day-age theory, but that view, too, is being abandoned for lack of support in the physical record.[64]

Concordism is problematic for other reasons as well. For one thing, no consensus has emerged about how the geological sequence of events is to be linked with the biblical sequence of events. Concordists have variously seen in Genesis 1:2, for example, references to a primitive globe, to the Earth under water, to universal matter, and to gaseous nebulae. They have variously linked the events of day two with the degassing of the Earth and the formation of galaxies or black holes. The Paleozoic Era has been equated with days two and three by some and

64. From time to time such periodicals as *Christianity Today* and *Eternity* publish articles on the relationship between geology and Scripture. As an example of such, see the October 1982 issue of *Christianity Today*. Among the more important popular books have been Davis A. Young, *Creation and the Flood* (Grand Rapids: Baker, 1977) and *Christianity and the Age of the Earth;* Robert C. Newman and Herman J. Eckelmann, *Genesis 1 and the Origin of the Earth* (Downers Grove, IL: InterVarsity, 1977); John Wiester, *The Genesis Connection* (Nashville: Thomas Nelson, 1983); Alan Hayward, *Creation and Evolution* (London: Triangle, 1985); and Michael R. Johnson, *Genesis, Geology and Catastrophism* (Exeter: Paternoster Press, 1988). And, of course, there is the voluminous scientific creationist literature that includes not only books, pamphlets, tracts, and newsletters but also films, slide-tape presentations, videotapes, and audio recordings.

with day five by others. Furthermore, concordists have always found it necessary to employ improbable exegesis to achieve a harmonization regarding day four (creation of the Sun, Moon, and stars) and days three and five (plants and animals). There is not one concordism; there are many *conflicting* concordisms, and not one among them does an adequate job of dealing with the diversity of questions raised by the evidence uncovered by biblical and scientific scholarship.[65]

From 1840 to the present, an overwhelming and diverse array of geological evidence has accumulated and helped to flesh out the outline of Earth history. This evidence has reinforced the perception of geologists, Christian and non-Christian alike, that the Earth is extremely old. For example, weathering and erosion of land surfaces are time-consuming processes. In thick stacks of layered sedimentary rocks there are generally several old erosional surfaces that have fossil soil zones and weathered rock beneath them. Each of these surfaces implies an extensive time of weathering.

The mathematical principles of heat flow are well understood. The rock record is replete with thousands of examples of large bodies of igneous rock that cooled from intensely hot liquids. Given the size, shape, and chemical composition of these bodies, we can calculate the approximate amount of time required for the liquid to have lost its heat and crystallized into rock. For many of the larger bodies of igneous rock the calculations indicate that thousands to millions of years may have been required.

Our knowledge of metamorphic rocks indicates that many of them were originally formed on the Earth's surface as sediments, then buried under extremely high pressures and subjected to severe temperatures. Even if a global flood could have buried thousands of feet of rock (which is highly unlikely), the amount of time that it would take for the thousands of cubic miles of rock to be heated to as much as 750°C and then to be recooled to surface temperature would be on the order of hundreds of thousands to millions of years. In addition, many layered rocks are strongly folded. Experimental studies of various rock types suggest that the times required to develop the observed folding in such rocks is on the order of tens of thousands of years or more.[66]

65. For an extensive documentation of the failures of various concordist approaches to the interpretation of Scripture, see Young, "Scripture in the Hands of Geologists."

66. For example, M. A. Biot, "Theory of Folding of Stratified Viscoelastic Media and Its Implications in Tectonics and Orogenesis," *Geological Society of America Bulletin* 72 (1961): 1595-1620.

Facies analysis has demonstrated again and again that layered sedimentary rocks were deposited in the kinds of surface environments with which we are familiar today.[67] For instance, there are examples of strata that formed in lakes, in deltas, in river valleys, as glacial deposits, as beaches, as coral reefs, as desert dunes, as alluvial fans at the base of mountains, and so on. The transition from one kind of environment to another may take thousands of years or more. Further, it takes long periods of time for any given environment to deposit considerable thicknesses of a particular rock type. Consider, for example, the slow rate at which lakes deposit bottom muds or the slow growth of coral reefs.

Finally, with the discovery of radioactivity, the development of a variety of radiometric dating techniques has permitted accurate and precise dating of thousands of rock specimens from around the world. Rocks and minerals are routinely found to have ages of tens to hundreds of millions of years. Radiometric techniques have also permitted calculation of the probable age of the Earth, currently accepted as about 4.55 billion years. This age agrees with ages that have been determined independently for the Moon and several meteorites that have fallen onto the Earth's surface.[68]

Because of the vast array of physical evidence, geologists today talk with warranted confidence of the age of the Earth, of oceans opening and closing, of continental land masses separating and colliding, of mountain chains arising and being eroded away, of life forms coming into existence and becoming extinct, and of the appearance and disappearance of glaciers, lakes, deserts, rivers, volcanoes, and other landforms. Geologists have even begun to work out the early history of the Moon and to correlate that history with terrestrial events.

So that the reader may gain a sense of the character and force of the data that have compelled students of the Earth to reckon with its antiquity, this chapter concludes with a brief survey of some of the geological features of one of the most spectacular regions of the world—the Colorado Plateau. The Colorado Plateau provides an outstanding example of why geologists long ago abandoned diluvialism and neptunism and accepted an ancient dynamic Earth.

67. Facies analysis is the study of sedimentary rocks with a view to reconstructing the original environment in which they were formed. Such analysis entails a variety of data, including mineralogy, texture, sedimentary structures, paleontology, associated rock types, and thickness.

68. See S. R. Taylor, *Planetary Science: A Lunar Perspective* (Houston: Lunar & Planetary Institute, 1982).

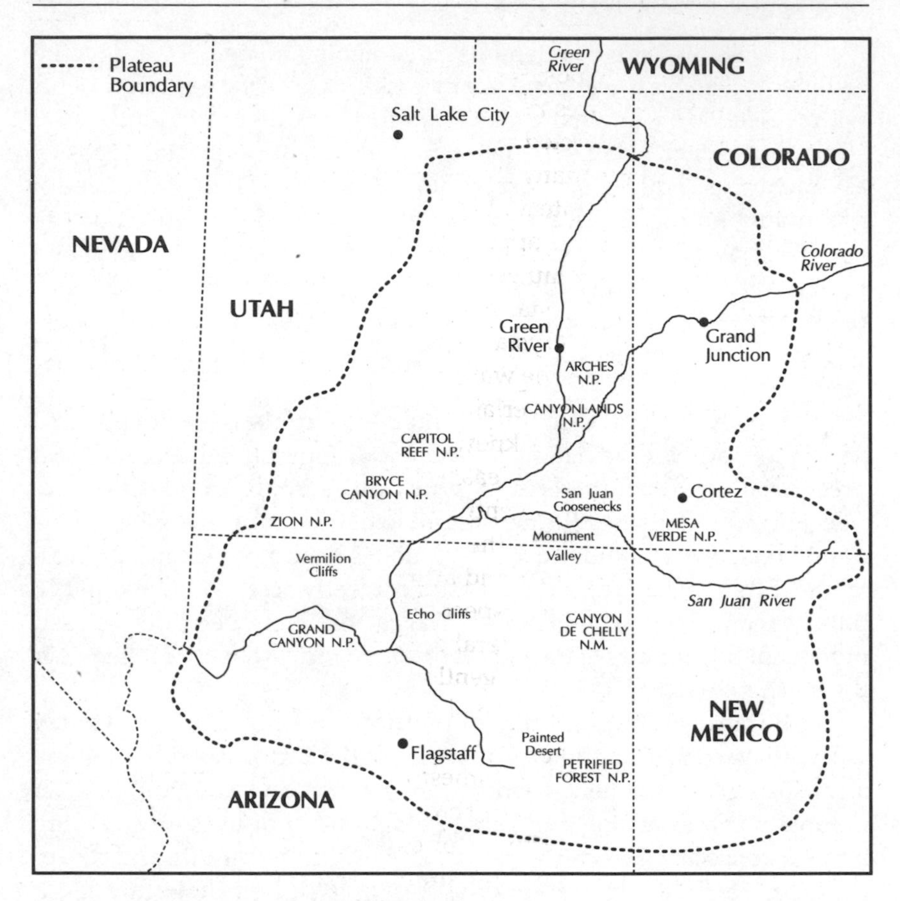

Figure 1. Location map of the Colorado Plateau

GEOLOGY OF THE COLORADO PLATEAU—A CASE STUDY

The basic geology of the Colorado Plateau is relatively simple and can be grasped even by the casual but observant traveler who has had little previous experience with geology (see Figure 1, above). The plateau consists of a very thick stack of layers of rock. The layers are nearly horizontal with some local tilting.[69] Examination of subsurface drilling

69. The total thickness of the rocks in the Colorado Plateau above the Precambrian basement varies from place to place but is on the order of 10-20,000 feet. The rocks are essentially horizontal with very broad-scale upwarped and downwarped folds superimposed on the basic horizontal pattern. The upwarps are called *domes,* and the downwarps are called *basins.*

samples and surface exposures indicates that the layers are generally continuous over distances of tens to hundreds of miles. Breaks in the continuity have been effected by erosion. The edges of the eroded layers are well exposed in the many cliffs, mesas, buttes, and canyons that are found throughout the plateau. Table 1, on page 64, lists the formation names that geologists have applied to layers of the Colorado Plateau.[70]

Our survey of plateau geology begins at the bottom of the Grand Canyon in northern Arizona. At canyon bottom, the Colorado River flows across a variety of crystalline rocks, including the Vishnu Schist and Zoroaster Granite. In the walls at the eastern end of the canyon the Vishnu and Zoroaster are overlain by about 12,000 feet of layered sedimentary and volcanic rocks known as the Unkar and Chuar Groups. These groups of layered rocks, easily visible from Lipan Point or Desert View on the south rim, dip gently toward the east. Above the Vishnu Schist, Zoroaster Granite, and the Unkar and Chuar Groups is a prominent angular unconformity,[71] and above that is 4,500 feet of horizontal rocks that make up most of the spectacular canyon walls (Figure 2, p. 65). The walls are composed of several distinct layers, some forming prominent cliffs and others forming gentler slopes.

The basal horizontal units are collectively known as the Tonto Group and consist of Tapeats Sandstone at the bottom, then Bright Angel Shale, and lastly Muav Limestone. The Tapeats is about 225 feet thick and is composed of resistant cross-bedded sandstone.[72] The Bright Angel consists of about 500 feet of easily weathered gray shale. The Muav is a 100-foot-thick cliff-forming limestone. Above the Tonto Group is the Redwall Limestone, a massive reddish cliff-forming layer about halfway up the canyon walls (Figure 3, p. 66). This cliff is composed of gray limestone that is stained red by iron oxides washing down from above. The Redwall is 500-800 feet thick and is rich in marine invertebrate remains. Above the Redwall are the Supai and Hermit

70. For an eminently readable and competent popular account of the geology of the Colorado Plateau that is full of photographs and maps and also features an extensive bibliography of technical geological sources, see Donald L. Baars, *The Colorado Plateau: A Geologic History* (Albuquerque: University of New Mexico Press, 1983). Readers who thoroughly familiarize themselves with this book will have a good understanding of the stratigraphy of the Grand Canyon and Colorado Plateau.

71. An "angular unconformity" is an old buried erosional land surface in which more nearly horizontal layered rocks are stacked atop more steeply tilted layered rocks.

72. Cross bedding is a phenomenon that occurs primarily in sandstones. It is characterized by layering or bedding of material at an angle to the primary bedding of the rock. Cross beds may be formed wherever a strong current moves sand, as in dunes, in riverbanks, and on beaches.

Table 1. Standard Formation Names Widely Used throughout the Colorado Plateau

Geologic Age	Name	
Tertiary	Green River Formation	
	Wasatch Formation	
Cretaceous	Mesa Verde Group	
	Mancos Shale	
	Dakota Sandstone	
Jurassic	Morisson Formation	
	San Rafael Group	Bluff Formation
		Summerville Formation
		Entrada Sandstone
		Carmel Formation
	Glen Canyon Group	Navaho Sandstone
Triassic		Kayenta Formation
		Wingate Sandstone
	Chinle Formation	
	Moenkopi Formation	
Permian	Kaibab Limestone	
	Toroweap Formation	
	Coconino Sandstone	
	Hermit Shale	
	Supai Group	
Pennsylvanian	Supai Group	
	Surprise Canyon Formation	
Mississippian	Redwall Limestone	
Devonian	Temple Butte Limestone	
Silurian		
Ordovician		
Cambrian	Tonto Group	Muav Limestone
		Bright Angel Shale
		Tapeats Sandstone
Precambrian	Chuar Group	
	Unkar Group	
	Vishnu Schist	

Formations, composed of several hundred feet of thin red shales, siltstones, and sandstones that contain ripple marks, raindrop impressions, fern imprints, and amphibian tracks. Separated from the Hermit by a sharp, smooth contact is the Coconino Sandstone, a very erosion-

Figure 2. North wall of the Grand Canyon as seen from Lipan Point. Toward the bottom of the canyon the Unkar and Chuar rocks dip gently to the right. An erosional surface (angular unconformity) separates them from overlying horizontal strata.

resistant cliff near the top of the Grand Canyon. The Coconino is nearly 400 feet thick and is composed of white, well-rounded, fine quartz sand grains whose surfaces are frosted and pitted.[73] The Coconino also contains spectacular cross beds (Figure 4, p. 67). Above the Coconino is the Toroweap Formation, a ledgy siltstone and limestone formation, and above that is the Kaibab Formation, the uppermost layer exposed in the walls of the Grand Canyon. The Kaibab, about 500 feet thick, is primarily a limestone that is rich in marine invertebrate fossils. North and east of the Grand Canyon, the Kaibab is overlain by great thicknesses of a variety of sandstones, mudstones, shales, siltstones, limestones, and evaporites, as indicated in Table 1.

Close examination of the features of the Grand Canyon and Colorado Plateau has provided compelling and coherent evidence concerning the ways in which the component rock formations came into existence and came to overlie one another as they have. It is the judgment of the entire professional geological community that diluvialist and neptunist models simply cannot account for a host of features of these

73. Frosting and pitting occur when sand grains are blown about by the wind. Frosting refers to the dull glaze that develops on a glassy surface, and pitting refers to tiny indentations in grain surfaces caused by constant collisions of sand grains with one another at high velocity.

Figure 3. Cliff of Redwall Limestone overlying more thinly layered Muav Limestone. Grand Canyon, Arizona

formations. In the pages that follow, we will take a look at some of this evidence and its implications.

1. Diluvialism entails belief that a flood occurred a few thousand years ago that destroyed all but those forms of life that were preserved on the ark. On a diluvialist model, the Unkar and Chuar Groups would be regarded as prediluvian and the Tonto Group would be regarded as the first deposits of the flood. If this were the case, we would expect to find the fossil remains of a wide variety of prediluvian animals and plants in those rocks formed immediately prior to the flood, and an even greater abundance in rocks formed immediately after the onset of the

Figure 4. Large-scale cross bedding in Coconino Sandstone, a probable desert sand dune deposit. Grand Canyon, Arizona.

flood, owing to the wholesale destruction of all species and their having been quickly buried in the sediments eroded by the surging waters. We ought, for example, to find an abundance of fossil trees, leaves, turtles, mice, large mammals, lizards, grasses, and perhaps even humans in the Tonto Group. In fact, however, the Unkar and Chuar Groups are devoid of any such fossil remains, while the Tonto Group contains only marine invertebrate fossils but absolutely no continental fossils at all. Indeed, of all the rock formations in the world that diluvialists have identified as "basal flood" deposits, not one contains the abundance of the continental fossils they theoretically ought to contain.

2. The rocks immediately beneath the angular unconformity between the Tapeats Sandstone and the underlying Unkar and Chuar Groups and the Vishnu Schist are thoroughly weathered and display local thick fossil soils that could not have survived catastrophic flooding.[74] The observed features indicate that the unconformity is an ancient land surface that experienced gentle weathering and erosion over a long period of time before being submerged beneath a gradually encroaching sea. The weathering and erosion effects could not have been produced by a catastrophic flood.

3. The succession of rocks within the Grand Canyon contains several other unconformities (erosional surfaces) that would not form in a global flood lasting one year. Formations from the Tapeats through the Muav represent marine deposits. The Muav is separated from the overlying Redwall by a distinct surface of erosion.[75] This erosional surface includes small channels that were scoured into the upper surface of the Muav and filled with Redwall sediments containing small pebbles of eroded Muav. These features indicate that before deposition of the Redwall, the Muav sediments had hardened into rock, risen above sea level, and been weathered to form pebbles and boulders that could be incorporated into the overlying sediments once the sea returned to the area. A global flood lasting only one year would have provided neither the time for the sediments to be consolidated nor opportunity for the materials to be weathered by exposure to the air.

Another important example of an unconformity is the contact between the Redwall Limestone and the overlying Supai Group. Observations by professional geologists indicate that the upper surface of the Redwall Limestone, although generally horizontal and conformable with the base of the overlying Supai, has many deep channels scoured into its upper surface—some as much as 400 feet deep. The channels are filled with layered mudstones, sandstones, and limestones and commonly contain pebbles derived from the Redwall.[76] These features indi-

74. Robert P. Sharp, "Ep-Archean and Ep-Algonkian Erosion Surfaces, Grand Canyon, Arizona," *Bulletin of the Geological Society of America* 51 (1940): 1235-70. Sharp provides excellent documentation of the abundant weathering and erosional effects that can be observed at the unconformity immediately beneath the Tapeats Sandstone. Such features would have been eliminated in any catastrophic flood.

75. The abundant evidence of physical erosion of material beneath the Redwall Limestone has been amply documented in the classic memoir of Edwin D. McKee and Raymond C. Gutschick, *History of the Redwall Limestone of Northern Arizona* (Boulder: Geological Society of America, 1969).

76. The abundant evidence for erosion between the Redwall and the Supai has

cate that the Redwall lime deposits were hardened into solid rock, lifted up from the seafloor to at least 400 feet above sea level, and there cut by flowing streams that dislodged pebbles from the exposed Redwall land surface and redeposited them in the channels. Still another indication that the Redwall was exposed to the atmosphere for a lengthy period of time—far more than a year—is the existence of caverns beneath, and of sinkholes in, its upper surface. The caverns and sinkholes are commonly filled with red shales from the overlying Supai Group or with angular blocks of fragmented Redwall. These features are virtually identical to the kinds of features observed today in the karst terrains of the southeastern United States.[77] The upper surface of the Redwall must have been exposed as land surface for a considerably long time to develop karst topography with sinkholes and caves.

Besides these unconformities between formations, several formations also have unconformities within them, including the Redwall Limestone and the Supai Group.[78] Each unconformity marks an extended period of the consolidation of underlying strata, uplift, weathering, erosion, and renewed sedimentation.

Besides these unconformities within the Grand Canyon, there are many other unconformities throughout the Colorado Plateau. For ex-

---

been superbly documented in both McKee and Gutschick, *History of the Redwall Limestone of Northern Arizona*, pp. 74-83, and Edwin D. McKee, "The Supai Group of Grand Canyon," U.S. Geological Survey Professional Paper 1173, 1982. Of particular importance in the latter work is the chapter on "Pre-Supai Buried Valleys," pp. 137-47. Included are some spectacular photographs of the deep channels that have been excavated into the upper surface of the Redwall Limestone. Recently it was determined that some of the deposits in the channels in the Redwall belong to a formation (since termed the Surprise Canyon Formation) that developed even before the Supai Group. See Stanley S. Beus, "A Geologic Surprise in the Grand Canyon," *Arizona Bureau of Geology and Mineral Technology Field Notes* 16 (1986): 1-4.

77. Karst topography generally develops in terrains that are underlain by soluble rocks, typically limestone. In the southeastern United States, for example, slightly acid groundwater slowly dissolves underground layers of limestone so that large fissure systems develop. Where the solution process continues long enough, many of the fissures develop into caves. Overlying roof rock may then sag or collapse entirely into these caves, resulting in surface depressions known as sinkholes or dolines. In karst terrains the landscape is typically dotted with hundreds of ovoid to circular pits of various sizes, many of them filled with ponds. The caves in the underlying limestone are filled with blocks of collapsed limestone and/or with the soils and sediments of overlying materials.

78. McKee and Gutschick, *History of the Redwall Limestone of Northern Arizona*, pp. 24-74, 155-76. The chapter on "erosion surfaces" documents thoroughly the evidence for intra-Supai unconformities as well as for the unconformity between the Supai Group and the overlying Hermit Shale. Included in the discussion are several excellent photographs of the unconformable contacts.

Figure 5. The Chinle Formation rests unconformably above gently tilted layers of the Moenkopi Formation. Small channels have been scoured into the unconformity between the two formations. Westwater Canyon of the Colorado River, Utah.

ample, a major unconformity separates the Kaibab Limestone, the uppermost layer exposed in the walls of of the Grand Canyon, from the overlying Moenkopi Formation, which crops out over wide areas of the plateau. Not only does the upper surface of the Kaibab have channels cut into it that are filled with Moenkopi rocks commonly containing pebbles of weathered Kaibab, but regional studies have indicated slight uplift and tilting of the consolidated rocks beneath the Moenkopi prior to Moenkopi deposition. There is also a major unconformity between the Moenkopi and the overlying Chinle Formation (Figure 5, above). This unconformity is marked by channels filled with Chinle sediments that locally contain pebbles of Kaibab and Moenkopi rocks. A prominent fossil soil zone characterizes the unconformity. There is an important unconformity between the Navajo Sandstone and the overlying Carmel/Entrada Sandstones (Figures 6 and 7, p. 71). The presence of each unconformity is physical evidence that the Colorado Plateau experienced consolidation of sediments, uplift and possibly gentle tilting, weathering of the uplifted surface to form soil, and erosion by streams and wind before the sediments of the next formation were deposited. There must have been several of these episodes of consolidation, uplift, weathering, and erosion—a conclusion clearly at variance with the

Figure 6. Layered Carmel Formation unconformably overlying massive Navajo Sandstone. Capitol Reef National Park, Utah.

Figure 7. Entrada Formation unconformably overlying Navajo Formation. Arches National Park, Utah.

theory that the sediments were deposited during a year-long global flood.[79] The numerous unconformities and alternations of marine and continental rocks are also at variance with the neptunist theory of sediment deposition from a slowly receding ocean.

4. The Coconino Sandstone contains spectacular cross bedding, vertebrate track fossils, and pitted and frosted sand grain surfaces. All these features are consistent with formation of the Coconino as desert sand dunes. The sandstone is composed almost entirely of quartz grains, and pure quartz sand does not form in floods. Quartz sand is developed only after very lengthy periods of intensive weathering and erosion of quartz-bearing rocks, such as granite, in which less resistant minerals such as feldspar and mica are destroyed, leaving only the relatively resistant quartz. Floods are too ephemeral to carry out the required weathering. So far as we know, deserts and beaches are the most likely environments for generating large concentrations of nearly pure quartz sand. In both cases a vast amount of time is required to weather away the less resistant minerals and concentrate the quartz. That a huge amount of rock had to be weathered and eroded in order to generate the quartz sand of the Coconino is evident from the fact that the Coconino forms a layer several hundred feet thick across most of northern Arizona.

The Coconino Sandstone is only one example of a desert dune deposit in the Colorado Plateau. Among the more prominent sandstones of the plateau are the Wingate Sandstone (Capitol Reef National Park and Colorado National Monument, Figure 8, p. 73), the Navajo Sandstone (Zion and Capitol Reef National Parks, Figure 9, p. 74), the Entrada Sandstone (Arches National Park, Figure 10, p. 75), the Bluff Sandstone (southeastern Utah), and the De Chelly Sandstone (Monument Valley and Canyon de Chelly National Monument, Figure 11, p. 76). Each of these very extensive formations is spectacularly cross bedded and composed almost entirely of frosted and pitted quartz sand grains. Like the Coconino, each of these sandstones required deposition of enormous

79. The general outlines of the evidence for the various episodes of sedimentation, uplift, erosion, and so on are summarized nicely in Baars, *The Colorado Plateau.* For further detail on the Kaibab-Moenkopi contact, see J. H. Stewart, F. G. Poole, and R. F. Wilson, "Stratigraphy and Origin of the Triassic Moenkopi Formation and Related Strata in the Colorado Plateau Region," U.S. Geological Survey Professional Paper 691, 1972, p. 15; on the Moenkopi-Chinle contact see J. H. Stewart, F. G. Poole, and R. F. Wilson, "Stratigraphy and Origin of the Chinle Formation and Related Upper Triassic Strata in the Colorado Pleateau Region," U.S. Geological Survey Professional Paper 690, pp. 14-19; on the contact of the Navajo with overlying formations, see J. W. Harshbarger, C. A. Repenning, and J. H. Irwin, "Stratigraphy of the Uppermost Triassic and the Jurassic Rocks of the Navajo Country," U.S. Geological Survey Professional Paper 291, 1957, p. 33.

Figure 8. Cliff of the massive Wingate Formation, a sandstone overlying the Chinle and Moenkopi Formations. Capitol Reef National Park, Utah.

volumes of thoroughly weathered and concentrated quartz sand under probable desert conditions. Given the total thickness of some of the formations, the amount of rock that had to be weathered and the amount of quartz that had to be concentrated is staggering. For example, at Zion National Park in southwestern Utah, the Navajo Sandstone is 2,000 feet thick, and in southeastern Utah it is still 200-300 feet thick. The Entrada Sandstone in eastern Utah is about 800 feet thick. The Wingate is about 300 feet thick. Each of these sandstones covers enormous areas of the plateau. No flood of any size could have produced such deposits of sand, and even if one could have produced the quartz, it could not have

Figure 9. Massive, spectacularly cross-bedded Navajo Sandstone, a dune deposit. Checkerboard Mesa, Zion National Park, Utah.

deposited such pure sand: the sand would necessarily have been contaminated with sediments brought in from other sources by the flood's turbulence.[80]

5. Many of the formations of the Colorado Plateau contain abundant fossil mudcracks. Mudcracks commonly develop on tidal flats or the shores of lakes when mud dries out. As the mud dries, it shrinks and cracks into individual plates that curl up with increased drying. Obviously, mudcracks could not have formed during flood conditions, but only afterward. The Supai Group within the Grand Canyon contains numerous layers with abundant mudcracks, as do the Moenkopi, the Chinle, and the Morrison Formations. Each of these formations had to experience several episodes of wetting and extended drying out. They cannot be global flood deposits.

80. For more detailed information on these sandstones, see Harshbarger, Repenning, and Irwin, "Stratigraphy of the Uppermost Triassic and the Jurassic Rocks of the Navajo Country," and Donald L. Baars, "Permian System of Colorado Plateau," *American Association of Petroleum Geologists Bulletin* 46 (1962): 149-218. For valuable discussions of eolian sandstones including reference to several of the sandstones of the Colorado Plateau, see Thomas S. Ahlbrandt and Steven G. Fryberger, "Introduction to Eolian Deposits," in *Sandstone Depositional Environments,* ed. Peter A. Scholle and Darwin Spearing (Tulsa: American Association of Petroleum Geologists, 1982), pp. 11-47; and Richard C. Selley, *Ancient Sedimentary Environments,* 3rd ed. (Ithaca: Cornell University Press, 1985), pp. 82-101.

Figure 10. Massive Entrada Formation, another dune deposit, overlying contorted beds of the Dewey Bridge Member of the Entrada. Arches National Park, Utah.

6. The Green River Formation occupies tens of thousands of square miles in the high plateaus of northwestern Colorado, southern Wyoming, and eastern Utah (Figure 12, p. 76). It averages 2,000 feet in thickness and in some places is as much as 6,000 feet thick. Much of the formation consists of oil shale containing organic material and fish fossils. The shale is composed of many extremely thin layers called *varves;* each varve consists of a very thin pair of laminations, one of which is carbonate-rich and the other of which is organic-rich. Many of the laminations are less

Figure 11. Cliff of de Chelly Sandstone, yet another desert dune deposit. Canyon de Chelly National Monument, Arizona.

Figure 12. Green River Formation. Many of the layers of this very thick unit were deposited in ancient long-lived lakes that covered large parts of Utah, Colorado, and Wyoming. Book Cliffs near Parachute, Colorado.

than one millimeter thick and yet can be traced laterally for several miles. There are more than a million vertically superimposed varve pairs in some parts of the Green River Formation. These varve deposits are almost certainly fossil lake-bottom sediments. If so, each pair of sediment layers represents an annual deposit.[81] Climatic changes during the year yield seasonally different sediment compositions. The lake deposits grade laterally into coarser sand and limy mudstone deposits on which have been found an abundance of mudcracks, bird tracks very similar to those of sandpipers and other web-footed birds, and skeletal remains of a flamingo-like bird. The Green River Formation also contains many layers rich in stromatolites—mound-like structures that are formed by gradual carbonate precipitation by algae. There are also layers rich in evaporite minerals. The sum total of the evidence indicates that the Green River sediments were deposited by huge, long-lasting, algae-rich lakes undergoing periodic evaporation. The mudcracked deposits containing bird tracks very probably represent muddy shores of the lake. The flamingo-like birds, whose remains have been found in abundance, are the kinds of birds that one would expect in great flocks along the shores of algae-rich lakes. Modern algae-eating flamingos live in flocks today along many of the lakes of the African Rift Valleys.[82] The fossil remains of fish appear to be freshwater species that died and sank to a stagnant, muddy, oxygen-poor lake bottom, where they were well preserved from scavengers. The total number of varve pairs indicates that the lakes existed for a few million years.

7. The plateau has been warped into great fold structures. Upturned strata are plainly visible at the eastern margin of the San Rafael Swell, at Capitol Reef (Figure 13, p. 78), and north of the Grand Canyon. Monument Valley lies at the eroded center of a very broad upwarp. The Grand Canyon cuts through strongly upwarped strata of the Kaibab and Coconino Plateaus. Regional analysis indicates that many of the large-

81. For a detailed discussion of the Green River varves, see Wilmot H. Bradley, "The Varves and Climate of the Green River Epoch," U.S. Geological Survey Professional Paper 158-E, 1929. There is voluminous literature on the Green River Formation. Among some of the more recent papers that include extensive bibliographies are George A. Desborough, "A Biogenic-Chemical Stratified Lake Model for the Origin of Oil Shale of the Green River Formation: An Alternative to the Playa-Lake Model," *Geological Society of America Bulletin* 89 (1978): 961-71; and Rex D. Cole and M. Dane Picard, "Comparative Mineralogy of Nearshore and Offshore Lacustrine Lithofacies, Parachute Creek Member of the Green River Formation, Piceance Creek Basin, Colorado, and Eastern Uinta Basin, Utah," *Geological Society of America Bulletin* 89 (1978): 1441-54.

82. For a popular presentation of the relationship between the fossil birds of the Green River Formation and their paleo-environment, see Alan Feduccia, "Presbyornis and the Evolution of Ducks and Flamingos," *American Scientist* 66 (1978): 298-304.

Figure 13. Tilted strata, indicating that large-scale warping and folding have been superimposed on the generally horizontal strata of the Colorado Plateau. Capitol Reef National Park, Utah.

Figure 14. Deeply incised or entrenched meanders at the Goosenecks of the San Juan River. The meandering pattern, similar to that of the Mississippi River, had already been established prior to very slow uplift of the region. Southeastern Utah.

scale folds developed after deposition of the Mesa Verde rocks and prior to deposition of Wasatch-Green River rocks or their equivalents, for in places Wasatch-Green River rocks or their equivalents rest unconformably on the partly eroded upturned edges of Mesa Verde rocks. Other upwarps are more recent than that. The upwarps cannot have developed catastrophically, or we would never see a feature like the Goosenecks of the San Juan River in southeastern Utah (Figure 14, p. 78). The intensely meandering pattern of the river could not have developed while the river cut down through the rocks, nor would it have developed just by following pre-existing fractures in the rocks. Rather, the meanders had to have been established prior to downcutting, as the river flowed across a flat surface like the present Mississippi River floodplain. The meander pattern could have been preserved only if the rate of uplift of the plateau (and therefore the rate of downcutting of the river while the plateau rose) was extremely slow. The meander pattern could not have been produced by flood waters rushing off the face of the Earth, nor could it have developed over a period of just a few thousand years. Nor is the San Juan River the only river to have cut a deep canyon through thousands of feet of layered rock. The Green River has cut impressive canyons through parts of eastern Utah, and in places the river has maintained a strongly meandering pattern that was established prior to downcutting. And, of course, the Colorado River has cut several canyons through western Colorado, eastern and southern Utah, and northern Arizona. The Grand Canyon is the most famous of these, but the Colorado has also cut the Debeque, Westwater, Cataract, Glen, and Marble Canyons, all impressive in their own right. These deep, locally sinuous gorges could not have been excavated by floodwaters.

8. There are many places in the Colorado Plateau where igneous rocks have intruded the layered rocks. Shiprock is an example of the deeply eroded remains of a volcanic pipe in northwestern New Mexico. The rock is composed of solidified lava that was injected into the Mancos Shale that underlies the local ground surface.

In southeastern Utah a few miles southeast of Capitol Reef are the Henry Mountains. These mountains consist of several bulging intrusions of igneous rock that were injected into the sedimentary rock layers of the area. The nature of the formation indicates that each individual body of rock must have cooled before the next batch of magma intruded, and each period of cooling is likely to have taken thousands of years.[83]

83. C. B. Hunt, P. Averitt, and R. L. Miller, "Geology and Geography of the Henry Mountains Region, Utah," U.S. Geological Survey Professional Paper 228, 1953.

Large quantities of magma were injected, leading to the arching of the overlying rocks of the Glen Canyon and San Rafael Groups (Table 1, p. 64). Subsequent erosion of the arched rocks has stripped away about 6,000-7,000 feet of overlying sedimentary rock and exposed many of the underlying bulges of igneous rock.

Volcanic activity has clearly affected the Grand Canyon area. One lava flow that is superimposed on all the Grand Canyon strata flowed down into the canyon and temporarily blocked the river. Radiometric dating indicates that this flow occurred 1.16 million years ago.[84] Among other things, this indicates that all the layered sedimentary rocks of the canyon must be older than 1.16 million years.

In summary, a substantial body of coherent evidence strongly suggests that the Colorado Plateau has undergone a very complex, dynamic history involving a variety of terrestrial environments and numerous interchanges of land and sea. Furthermore, the Colorado Plateau is only one example of such complexity. Similar reconstructions of geological history can be produced from evidence uncovered about the Appalachians, the Alps, the Canadian Rockies, the Andes, the British Isles, southern Africa, or anywhere else in the world. The histories recorded by the rocks in these places must be long ones, for everything in our experience tells us that it takes a long time for land and sea to interchange on a large scale and for one kind of environment to be succeeded by another. It takes time for sediments to harden to rock, for karst topography to develop, for erosion to occur, for soil zones to develop, for rivers to deposit great thicknesses of sediment, for magma to crystallize far underground, for rocks to be tilted, for lakes to accumulate millions of varves, for rocks to be folded, for rivers to cut deep canyons, and so on. The totality of the evidence points to a time frame that is vastly greater than that of recorded human history. Mountains of evidence indicate that the Earth is very old.

If rocks are historical documents, we are driven to the related conclusion that the available evidence is overwhelmingly opposed to the notion that the Noahic flood deposited rocks of the Colorado Plateau only a few thousand years ago or that the rocks were formed from a diminishing ocean. The global deluge hypothesis fails to account for fossil mudcracks, soil zones, unconformities, pure quartz sand deposits, frosted and pitted quartz grains, thinly varved muds, karst topography, lithification, the distribution of terrestrial fossils, folds, igneous intru-

84. E. D. McKee, W. K. Hamblin, and P. E. Damon, "K-Ar Age of Lava Dam in Grand Canyon," *Geological Society of America Bulletin* 79 (1968): 133-36.

sions, and many other features. The neptunist hypothesis fails to account for the alternations of marine and terrestrial deposits and for the several episodes of uplift indicated by the unconformities. The Christian who believes that the idea of an ancient Earth is unbiblical would do better to deny the validity of any kind of historical geology and insist that the rocks must be the product of pure miracle rather than try to explain them in terms of the flood. An examination of the Earth apart from ideological presuppositions is bound to lead to the conclusion that it is ancient.

4. THE SCIENTIFIC INVESTIGATION OF COSMIC HISTORY

HOWARD J. VAN TILL

The discovery of terrestrial history was soon followed by the discovery of history in the world of stars and galaxies. How is such a discovery possible? How can reliable information about distant celestial objects be obtained? How can astronomers formulate credible theories concerning the behavior and formative history of things too remote to visit? These questions and the answers offered by contemporary astronomy and cosmology are discussed in this chapter by Howard J. Van Till.

But what about these scientific answers? How should the Christian community assess the concept of fifteen billion years of cosmic evolution proceeding from a big-bang beginning? Is the scientifically informed evolutionary world picture consonant with a biblically informed Christian worldview? This chapter closes with some critical reflection on these matters and calls for continuing discussion in which important distinctions between scientific and theological matters are honored.

CELESTIAL LUMINARIES: FROM DIVINE BEINGS TO MATERIAL BODIES

IN MANY ANCIENT Near Eastern cultures, stars and other celestial luminaries were not thought of as being material entities comparable to ordinary terrestrial objects; rather, they were commonly perceived to be celestial deities.[1] Whether they were seen as gods themselves or as the visible manifestations of astral deities, they came to be viewed as sentient, volitional beings with influence over human affairs. Born in Mesopotamian polytheism and nurtured in both Greek and Roman culture, astrology became a deeply entrenched cultural phenomenon—one that persists to this day.[2]

In medieval cosmology, stars were no longer consistently equated with divine beings, but they were still considered entities that were essentially different from things in the terrestrial realm. The earthly environment was thought to be composed of four elements—air, earth, water, and fire—but the stars were thought to be made of a fifth kind of substance: the "quintessence." As such, they could be admired for their many perfections, as exhibited in the regularity of their motions, the constancy of their brightness, and the like. However, to perceive them or to study them as ordinary physical objects, bound to the same principles of material behavior as terrestrial phenomena, would have been unthinkable to the medieval mind. The celestial and terrestrial realms were considered to be distinct and fundamentally different from one another.

As children of the twentieth century who have grown up in the context of modern natural science, we approach the investigation of stars in a manner very different from the ancient or medieval person. We perceive the stars as material bodies made of the same kinds of elements that we find here on Earth and behaving according to the same patterns of material behavior that we see exhibited in terrestrial laboratories. Though the stars are unimaginably distant from us, we now have instruments capable of gathering light from individual stars and harvesting a wealth of information concerning their physical properties and the physical processes occurring near their surfaces. This information, coupled with theoretical models for the structure and behavior of stellar interiors, has opened the way to a detailed and computationally

1. For a helpful introduction to the ancient Near Eastern perspective, see Henri Frankfort et al., *The Intellectual Adventure of Ancient Man* (Chicago: University of Chicago Press, 1946).

2. On this, see Franz Cumont, *Astrology and Religion among the Greeks and Romans* (1912; reprint ed., New York: Dover Publications, 1960).

sophisticated investigation of the temporal development (or "life history") of stars and stellar systems. In this chapter we will briefly describe the results of such observational and theoretical investigation, and we will discuss some of the aspects of that enterprise that appear to be problematic to the Christian community.

STARS ARE NOT INDEPENDENT, isolated entities; they are members of interacting systems of celestial bodies. The Sun, for example, a rather ordinary star, is the principal member of our solar system—the family of planets, meteoroids, comets, and asteroids moving in gravitationally bound orbits around the Sun. In the previous chapter, Davis Young reviewed the development of a scientific reconstruction of Earth history drawn from the physical evidence left by geological events and processes. Is the biography of the Sun (and other stars) consistent with this terrestrial history? Are the time scales comparable? Is there evidence that stellar history and planetary history are coherently interrelated?

And what if we look at the Sun in relation to other stars? The Sun is but one star among several hundred billion stars that are arranged in the gigantic, spiral-armed stellar system known as the Milky Way Galaxy. Does this galaxy, a hundred thousand light-years in diameter, have a discernible history? If so, is galactic history coherently related to stellar history?

And what about the rest of the incomprehensibly vast universe that is now accessible to observation through the telescopes of contemporary astronomy? The visible universe is made up of tens or hundreds of billions of galaxies, most of them similar to the Milky Way, distributed throughout an expanding spatial domain with dimensions measured in many billions of light-years. Is it possible to investigate the history of the entire visible cosmos? What clues has cosmic history left to be uncovered? Do planetary history, stellar history, and galactic history fit comfortably within the framework of cosmic history? Does the temporal development of the entire cosmos exhibit any measure of coherence? Of directionality? Of purpose?

These are fascinating and important questions. We are privileged to live at a time when we are able to carry out a search for credible answers to such questions drawn from an investigation of the universe itself. And because the universe that we study is God's Creation, we proceed in the confidence that the structures, processes, and historical records that it exhibits will be trustworthy manifestations of the handiwork of God.

STARS: EVOLVING THERMONUCLEAR FURNACES

Starlight as Information

How do we know of the existence of stars? Their presence is revealed by their luminosity. One upward glance on a clear, moonless night informs us of the existence of thousands of luminous stars. We know stars exist because we see the light they radiate.

But starlight reveals far more than the simple fact that stars exist. The light radiated by stars is permeated with information about their physical properties. From both the intensity and wavelength distribution of starlight we may learn the value of a star's surface temperature, its chemical composition, its luminosity, its distance, its motions, its size, and several other aspects of its physical nature and behavior.

Our concern here is not to provide the technical instruction needed to perform the collection and analysis of starlight that a professional astronomer performs. Those who are interested in these fascinating details should consult the technical and textbook literature that is readily available at a variety of levels.[3] Here we wish only to report representative results of astronomical investigation so that the reader may become familiar with some of the information that can be gleaned from the study of starlight—information that yields much knowledge about the properties, behavior, and history of the Creation.

May such information be ignored? Would closing one's eyes to the knowledge that can be gained by a scientific investigation of God's universe be honoring the Creator? We think not. While it is obvious that not every Christian is called to be engaged in a formal study of the sciences, those of us who are involved in any of the several aspects of Christian education—as teachers, pastors, or parents—are called to stimulate the community of believers to explore God's handiwork by all legitimate means. Astronomy, along with all of the other natural sciences, is one of the tools that is available to serve as an aid in that exploration.

3. For a descriptive introduction to contemporary astronomy, see Michael A. Seeds, *Horizons: Exploring the Universe,* 3rd ed. (Belmont, CA: Wadsworth Publishing, 1989), or William J. Kaufmann III, *Discovering the Universe,* 2nd ed. (New York: W. H. Freeman, 1990). For a somewhat more quantitative overview, see Michael A. Seeds, *Foundations of Astronomy,* 2nd ed. (Belmont, CA: Wadsworth Publishing, 1988), or William J. Kaufmann III, *Universe,* 2nd ed. (New York: W. H. Freeman, 1988). For an introductory text that includes more astrophysical detail and deals candidly with unsolved puzzles in contemporary astronomy, see Frank H. Shu's *The Physical Universe* (Mill Valley, CA: University Science Books, 1982).

The Several Families of Stars

An empirical (i.e., observational) study of the physical properties of stars soon reveals that stars are not identical to one another. They differ in a number of significant ways. It is of great interest to astronomers to compare stars on the basis of selected properties.

Early in the twentieth century, Ejnar Hertzsprung and Henry Norris Russell discovered a particularly fruitful approach to such a comparative study. Describing their work in modern terms, we say that Hertzsprung and Russell chose to compare stars on the basis of their surface temperature and luminosity values. A star's surface temperature can be determined from the wavelength spectrum, or color composition, of its light. Just as the color of the light emitted by an electric heating element depends upon its surface temperature, so the temperature of a star's surface is revealed by the color of starlight. Stellar surface temperatures are found to lie within the range of about 2,500K to about 50,000K.[4] The Sun's surface is maintained at a temperature of approximately 5,800K.

The "luminosity" of a star is defined to be the rate at which it is radiating energy in the form of light and other electromagnetic radiation. Its value may be expressed in watts, just as we do for ordinary light bulbs. The determination of stellar luminosity is ordinarily based on the measured values of a star's distance and brightness. Luminosity values determined in this manner span a broad range—from approximately 0.0001 to 1,000,000 times the luminosity of the Sun.

The observation that stars differ from one another in surface temperature and luminosity is by itself not particularly significant. What is very remarkable, however, is the particular pattern of differences that stars exhibit. The best way to illustrate that pattern is to construct a diagram on which stellar luminosity values are plotted along one axis of the diagram and stellar surface temperatures are plotted on the other. Following the convention established by Hertzsprung and Russell, Figure 1 shows the result of placing on a luminosity-temperature diagram (or H-R diagram, in honor of Hertzsprung and Russell) points that represent the combination of luminosity and temperature values for a large number of representative stars.

4. All temperature values will be expressed in Kelvins, K. On the Kelvin temperature scale, 0K is the absolute zero temperature, and the freezing and boiling points of water are 273K and 373K respectively.

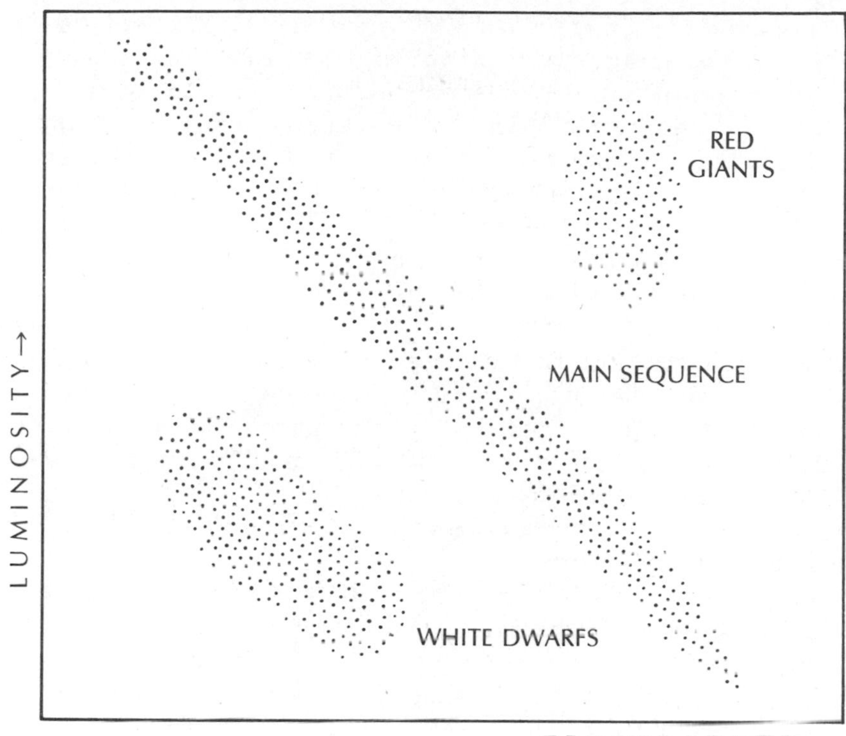

Figure 1. A Representative Hertzsprung-Russell Diagram. The points on this H-R diagram represent stars positioned according to their luminosity and temperature values. The vertical axis marks luminosity values increasing from bottom to top, and the horizontal axis marks surface temperatures increasing from right to left (the hotter the stellar surface, the farther to the left it will be placed on the diagram). This simplified diagram shows three of the most common families into which observed stars typically fall when charted in this fashion.

An inspection of Figure 1 immediately reveals that certain combinations of luminosity and temperature are more commonly found than others. The points representing stars are not randomly scattered all over the diagram; rather, they are most likely to be found in one of three major regions, labeled "main sequence," "red giant," and "white dwarf."

About ninety percent of all stars are main-sequence stars, forming a band that runs diagonally across the H-R diagram. Further measurements reveal that the stars comprising this band are also distributed along the main sequence according to the value of their mass (a quantity related to weight).[5] Stars above and to the left of the Sun's position on the main sequence have greater mass values. Moving down and to the right along the main sequence, we encounter progressively less massive stars.

In the upper right portion of the diagram we find stars that are radically different from main-sequence stars. Although they have mass values comparable to main-sequence stars, these specimens have relatively lower surface temperatures, giving them a distinctively more reddish color. But their size sets them apart even more radically: the typical diameter of one of these stars is a few hundred million miles, hundreds of times larger than most main-sequence stars. We appropriately call these stars "red giants." Antares, the prominent red giant star in the constellation Scorpius, has a diameter of about 600 million miles. If the Sun were replaced by Antares, planet Earth would reside within the interior of the stellar surface; we would soon be reduced to little more than a wisp of tenuous stellar gas.

The opposite extreme is found in the lower left corner of the H-R diagram. The stars represented by points clustered in this region are relatively high in surface temperature, thus displaying a color best described as "white." As revealed by their very low values of luminosity, these stars are remarkably small. Just as the red giants were found to be a few hundred times larger than main-sequence stars, so these "white dwarfs" are found to be a hundred times smaller. A typical white dwarf star has a diameter of about ten thousand miles, comparable in size to planet Earth.

This project of comparing stars on the basis of their luminosity and surface temperature values has led to a remarkable discovery: stars belong to distinctively different families; stars, like flowers, come in very definite varieties. Stars are not just stars; they are main-sequence stars, or red giant stars, or white dwarf stars, or members of some other specific category of stars that displays a combination of properties peculiar to

5. The term *mass* should not be confused with either *volume* or *weight*.

Newton's second law of motion indicates that if a net force is applied to an object, it will respond by accelerating. The mass of an object can be computed by using the relation

mass = net force applied/resulting acceleration

The greater the mass, the smaller the acceleration resulting from a given applied force. Mass, therefore, is a measure of the ability of an object to resist acceleration under the action of a net applied force.

It is also the case that the *weight* of an object near the surface of the Earth (that is, the strength of the force of gravity exerted by the Earth on that object) is directly proportional to its mass. Specifically, weight = mass × acceleration due to gravity. Thus, mass and weight are related, but are not the same kind of quantity.

that category. Like people, stars exhibit recognizable family traits. This discovery that stars belong to families immediately raises a number of interesting questions that deserve further investigation. Is it possible to identify the physical phenomena that cause these stellar families to differ from one another? Why, for instance, is one star a main-sequence star while another is a red giant? What causes dwarf stars to be so small compared with giant stars? Why are main-sequence stars distributed along a diagonal band on the H-R diagram according to their mass values? How are these families of stars related to one another? The patterned relationship among stellar properties revealed by the H-R diagram provides the occasion and stimulus to ask an extended series of more specific questions about the nature and behavior of stars. But to find answers to these questions we must first deal with the basic question concerning the generation of starlight.

Why Do Stars Shine?

An upward glance on a clear night reveals the existence of thousands of scintillating points of light. With the aid of a telescope, the number of visible stars multiplies to countless millions. Each star reveals its presence by the light that it radiates. But what leads stars to produce light? What physical process generates the energy required to maintain the high temperature of a luminous stellar surface?

Could the surface of a star be on fire? A view of the Sun at sunset might suggest that it is a continuously burning sphere. But that cannot be the case. Three-fourths of the Sun's mass is in the form of hydrogen, an excellent fuel, but there is no oxygen supply.[6] If no oxygen, then no fire; the rules on the solar surface are the same as those on Earth.[7] Though the Sun's surface is as hot as a fire, its high temperature must be maintained by some other process.

In the mid-nineteenth century, Hermann von Helmholtz computationally investigated another process that could supply the energy to heat the surface of the Sun and other stars—the process of gravitational

6. Information concerning the chemical composition of stars can be derived from the wavelength spectrum of starlight. For an introduction to the principles and methodology of extracting such information from starlight, consult any general astronomy textbook, such as George O. Abell, *Exploration of the Universe,* 4th ed. (New York: Saunders College Publishing, 1982), pp. 423-33.

7. We are here making the standard working assumption known generally as the "principle of uniformity." Though we call it an assumption, it is supported by a vast array of evidence and its employment characterizes the contemporary scientific enterprise.

contraction. The ball of gas that constitutes a star is held together by gravity. Each atom is gravitationally attracted to the rest of the star's material, resulting in a spherically shaped ball held together by its own weight. Assuming that all of the ordinary laws of physics (the patterns describing physical behavior in relationship to material properties) apply to stellar matter, Helmholtz determined that a star should gradually contract. This contraction would lead to the conversion of gravitational potential energy into heat, thereby providing an energy supply for a star's luminosity.

But Helmholtz contraction alone cannot be the complete answer to our question concerning the maintenance of stellar luminosity.[8] So, as in the construction of any adequate scientific theory, we must inquire concerning the possible roles played by other relevant phenomena. In this case we must explore the phenomenon of thermonuclear fusion.

If a cloud of gas collapses and contracts under the action of gravity, it becomes hot. Not only does the surface become as hot as the observed surfaces of stars, but the more highly compressed interior becomes even hotter. According to computations based on the thermodynamic behavior of gases, the center of a star achieves temperatures of more than 10 million K. At such astronomically high temperatures, the process of thermonuclear fusion takes place. This process, first investigated in the 1930s, generates an enormous quantity of energy as a by-product of the fusion of small atomic nuclei to form larger ones. (At the present time, we have learned to use this energy only destructively, in the hydrogen bomb; research into constructive uses continues.) Under the conditions that prevail in the solar interior, most of the Sun's energy is being generated by a series of thermonuclear fusion reactions which consume hydrogen to produce helium. This energy slowly works its way out from the intensely hot core region to the Sun's surface, providing heat at precisely the correct rate to maintain the steady surface temperature and luminosity that we observe and depend on for our terrestrial comfort.

We often take the daily supply of sunshine and the nightly display of starlight for granted. These familiar phenomena, however, are the direct consequence of physical processes that are continuously occurring. Without the continuing occurrence of thermonuclear fusion in the setting provided by gravitational compression, stars would not shine;

8. One of the standard arguments in support of this statement is that the contraction process is unable to supply energy for the multibillion-year duration of the Sun's history. But other arguments, quite independent of time scale considerations, are equally supportive. For example, Helmholtz contraction cannot account for the principal features of the H-R diagram in a consistent fashion.

the world would be a cold and dark place—nothing like the world whose comforts and pleasures we daily enjoy.

THE LIFE HISTORY OF STARS

Night after night the stars appear to the casual observer as the epitome of constancy and changelessness. Today, however, stars are interesting not by virtue of their apparent constancy, but because they are undergoing a process of persistent and irreversible change. Though these changes may occur at an imperceptibly slow rate (by human standards), their cumulative effect is of great significance.

Stars must change. Change is a necessary part of a star's existence as a luminous object, just as change is a necessary aspect of a bonfire. As we have seen, stars can be luminous only at the expense of the transformation of some stored energy supply into the energy that they radiate as light. That transformation process is necessarily a process of change, whether it is a chemical process, a gravitational process, a nuclear process, or some other physical process. While the total *amount* of energy is conserved, one *form* of energy is being changed into another form.[9] What remains after the process is different from what was present at the beginning, just as the ashes of a bonfire on the beach are different from a pile of driftwood.

On the basis of numerous clues drawn from starlight, astronomers are confident that the energy-producing processes that occur in stars conform to the same patterns of material behavior as those that have been observed in terrestrial laboratories. The light radiated by stars can be fully explained in terms of the behavior of a hot stellar surface surrounded by an atmospheric blanket of cooler gases. Light coming to us from stars and even more distant objects strongly suggests to us that the patterns for material behavior are spatially and temporally invariant—the same at places and times very remote from here and now. To say that the behavior patterns of the universe are uniform and unchanging is, in the judgment of the community of professional scientists, to make a statement that honestly reflects our experience with the material world.

Having recognized this fruitful presupposition as a cornerstone

9. We are here making the standard assumption that the principle of energy conservation applies. In *all* physical phenomena so far examined, the total amount of energy contained by a system plus its environment remains constant despite changes in either the form or location of that energy. Energy, we therefore say, is a conserved quantity.

of modern natural science, we may be led to ask why it should be so. Why should this universe behave in accordance with invariant patterns? Such a question stands at a higher level than ordinary scientific questions concerning the intrinsic intelligibility of the world we experience.[10] At this "metascientific" level, a person's philosophical, theological, and religious perspectives come to the fore. While a philosophical naturalist may seek to understand the invariance of physical law as a necessary consequence of autonomous material behavior, a Christian perceives the same behavior patterns as evidence for faithful and consistent patterns of divine governance. As Creator of the cosmos, God sustains and governs his Creation in a dependable and coherent manner; the concept of an undependable, capricious, or incoherent creator is wholly inconsistent with the Creator revealed in Scripture.

Having made a brief excursion into the realm of metascience, let us return to our consideration of the life history of stars, a topic within the restricted domain of natural science. Working in the context of the presupposition that the behavior and history of stars are characterized by a continuity of coherently patterned processes and events, astronomers have been able, particularly within the last few decades, to reconstruct the pattern of stellar life history. It is a fascinating story. To most persons it is a story filled with surprises; having inherited the concept that stars are paragons of constancy, we may be startled to discover that they have histories, that they are born, grow old, and die. Perhaps even more surprising, these histories can be systematically investigated.

On the basis of a wealth of evidence, interpreted according to the principles outlined above, astronomers have reached the conclusion that stars are born by gravitational collapse. Within numerous large interstellar nebulae, massive globules of gas and cosmic dust may, in the course of their response to environmental forces, satisfy the conditions for gravitational collapse.[11] This gravity-induced process heats the gas as it

10. We will consider the idea of "intrinsic intelligibility" more fully in the next chapter. In essence, it denotes that feature of the physical universe that allows us to come to some knowledge and understanding of its properties, behavior, and formative history (along with their interrelationships) on the basis of the empirical/theoretical techniques of contemporary natural science.

11. An *interstellar nebula* is a large cloud of mostly gaseous material found most commonly within the spiral arm regions of a galaxy. Within these nebulae, which usually have dimensions of many dozens of light-years and contain an amount of material comparable to thousands of stars, are smaller concentrations of material. If these "globules" have sufficient mass contained within a small enough volume, the gravitational attraction of each part for the rest may be strong enough to lead the globule to shrink to a star-size object. This is the process we are calling *gravitational collapse.*

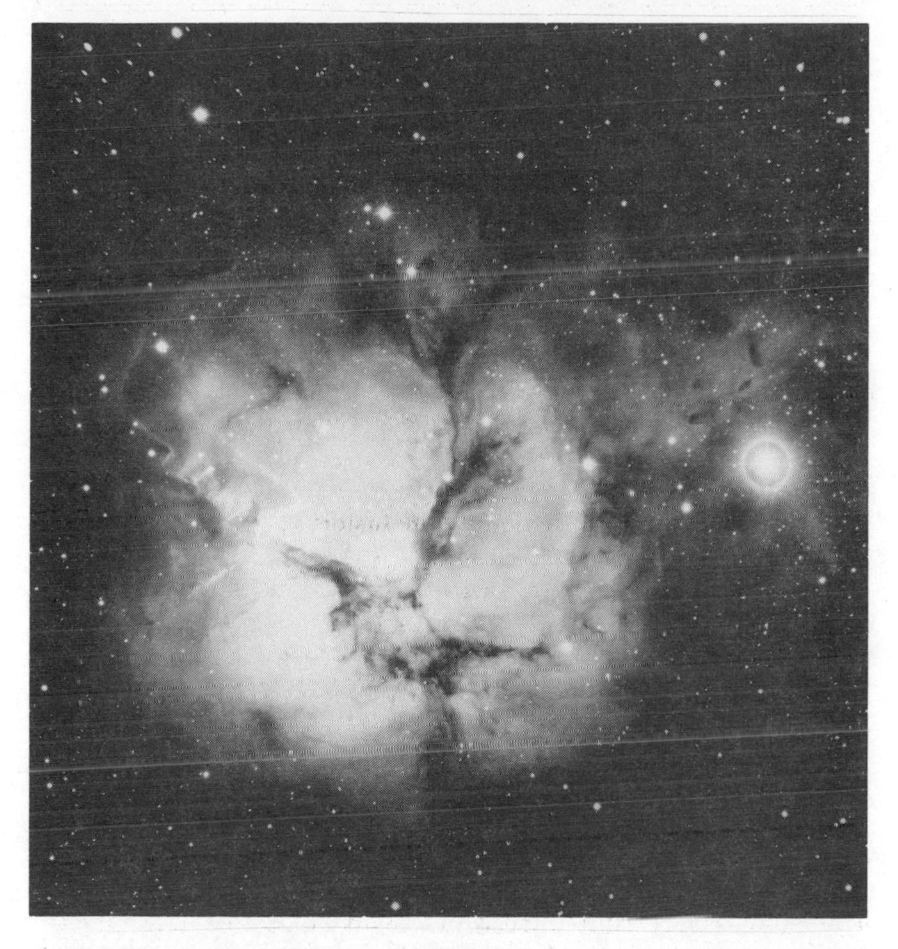

The Trifid Nebula, in the constellation Sagittarius, is one example of an interstellar nebula in which star formation is taking place. For additional examples, see pp. 156-58.
Palomar Observatory Photograph

is compressed into a progressively smaller volume. Recent infrared observations have revealed the presence of a large number of "hot spots" within nebulae—just the sort of thing one would expect if starbirth is occurring at the present time. As the gravitational contraction proceeds, the temperature of the protostar core is elevated to sufficiently high temperatures to ignite the thermonuclear fusion reaction. The energy generated by the fusion process produces an outward-directed thermal pressure that brings contraction to a halt and establishes an equilibrium state. The newborn star is now a relatively stable object.

But what kind of star is it? To what family of stars does it belong?

According to computations that seek to determine the properties of objects formed according to the sequence of events and processes described above, the location of a newborn star on an H-R diagram is fixed by the amount of its mass. The most massive stars, characterized by high values of surface temperature and luminosity, will be found in the upper left corner. Less massive stars will be distributed sequentially along a band running diagonally toward the lower right corner. In other words, the *computed* behavior of newborn stars stabilized by the thermonuclear fusion of hydrogen into helium duplicates the *observed* behavior of main-sequence stars. This correspondence is not contrived; it is the direct consequence of our previously stated assumption concerning the universal and invariant character of the laws of physics. Having made that presupposition, and having computed its consequences for the properties of stars, we are able to recognize main-sequence stars as stars that are in the first stable period of their lifetime.

During the main-sequence phase, thermonuclear fusion slowly converts the star's hydrogen fuel into helium. As a consequence of this internal change in chemical composition, a main-sequence star slowly ascends along a vertical path on the H-R diagram, traversing the width of the main-sequence region. Eventually, the cumulative effects of helium production lead the star to increase greatly in both size and luminosity while diminishing in surface temperature. The product of this transformation is easily recognized as a red giant star, found in the upper right portion of the H-R diagram.

The time scale for the processes and events so far described is astounding. For a protostar with a mass comparable to the Sun, the birth process—from the onset of gravitational collapse to the establishment of equilibrium between thermal pressure and gravity—lasts approximately 30 million years.[12] The main-sequence phase represents the stable adult phase of a star's lifetime; a star like the Sun is computed to remain in this phase for 10 billion years. More massive stars move through their birth and maturation processes more quickly; less massive stars mature more slowly. The Sun, with an age of 4.6 billion years,[13] appears to be a middle-aged star, roughly halfway between birth and death.

12. This number, like many of the others quoted in this overview, is based on computational models for the structure and behavior of stars—models that are constrained to conform to all of the laws of physics presently known.

13. Because of circumstantial evidence that planets are a by-product of star birth, and because the age of Earth, Moon, and planets has been determined to be 4.6 billion years, we use this same value for the age of the Sun.

Upon entering the red giant phase, stars proceed more rapidly through a sequence of episodes whose particular features are fixed by the mass value of the original star. Some red giants shed their outer envelope and leave behind a small core which can no longer support the fusion process. This dense stellar remnant is none other than the white dwarf which we introduced earlier. More massive stars approach the end of their lifetime in a more dramatic way by exploding as a "supernova." The most massive stars formed may omit the supernova explosion and experience a catastrophic collapse into a "black hole." This bizarre stellar corpse represents the ultimate victory of gravity over all opponents, concentrating mass into so small a volume that the resulting gravitational forces within a few mile radius are so intense that nothing can escape from that sphere of distorted space-time, not even light.

While many fascinating details have been omitted from our discussion of stellar history, note what we have discovered: the principal categories of stars—main-sequence, red giant, and white dwarf—are related to one another as members of a temporal sequence. Ordinary physical processes such as gravitational collapse, compressional heating, thermonuclear fusion, and the radiation of light from a heated surface have brought a cold globule of interstellar gas through a succession of physical processes and events that constitute the life history of a star. Gravitational collapse leads to the formation of a main-sequence star; the effects of thermonuclear fusion eventually transform a main-sequence star into a red giant; subsequent physical processes within a typical red giant cause it to shed its outer envelope and leave behind a white dwarf star. The families of stars that we discovered on the H-R diagram may now be recognized as stages in stellar development.

The succession of physical processes and events that we have been describing is what astronomers call "stellar evolution." In terms of our present understanding, stellar evolution is the inevitable consequence of the manner in which the behavior of matter is governed. Unless we postulate that stars may violate the rules for the behavior of material systems, we must grant that stars undergo change and development as a consequence of the energy generation processes that maintain their luminosity. The consequence of stellar behavior is the evolution of stellar properties. Fully conforming to the ordinary patterns for material behavior, stellar evolution represents the normal, irreversible, and cumulative consequence of an enormous number of individual events and processes at the atomic level. To avoid this directional flow of events from stellar birth to stellar death would require an entirely

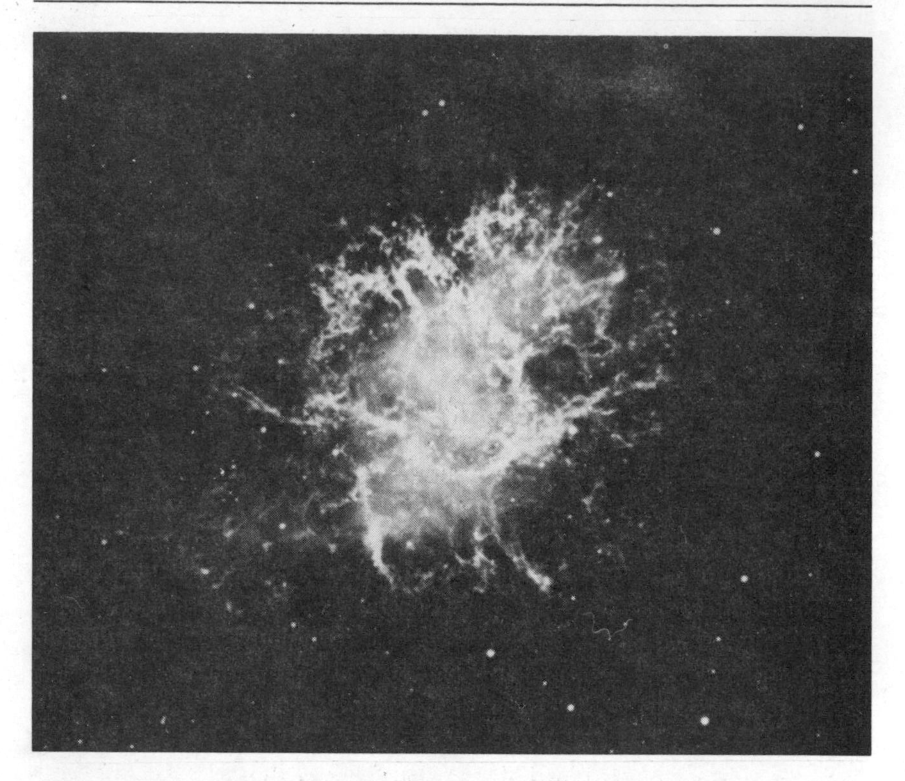

The Crab Nebula, remains of a supernova explosion observed in A.D. 1054.

Palomar Observatory Photograph

different set of rules for material behavior—rules for a world remarkably different from the one in which we live.

Stellar evolution, we now see, is no extraordinary or exceptional phenomenon; it requires no circumvention of the laws for gravity, nuclear reactions, or thermodynamics. Stellar evolution is neither more nor less than the cumulative consequence of ordinary material behavior; it is the product of processes just like the ones we observe every moment of our own lives. To deny, therefore, that stellar evolution is now taking place is to demand that the universe behave incoherently, that stellar material behave according to a set of patterns inconsistent with the ones we ordinarily observe. And to deny that stellar evolution has occurred throughout a multibillion-year history is to demand that stars appear to have experienced a history very different from their actual history. It is to demand that the evidence for their history is a mere illusion, that their apparent history is not authentic history.

GALAXIES: HERALDS OF COSMIC HISTORY

The Cosmic Distance Scale

We are explorers of space, we say. We have sent men to the Moon, we have landed measuring instruments on the planets Mars and Venus, and we have photographed the outer planets from passing spacecraft. These are impressive technological feats, to be sure, and yet to claim them as warrant for calling ourselves space explorers sounds to an astronomer a bit like a child claiming to be an Earth explorer on the basis of his brief excursion into the backyard.

It's a small world, some people say. Don't believe it. We live in an unimaginably vast universe. The Earth is nearly a hundred million miles from the Sun. Pluto, the most distant of the known planets in our solar system, maintains an average distance from the Sun of about four billion miles. Even these solar system dimensions, however, are minuscule compared with others that we encounter. The nearest star—we call it Proxima Centauri—is more than six thousand times farther away from us than Pluto, a distance of approximately 25 trillion miles.

To aid us in comprehending the vast scale of cosmic distances, we need a conveniently large unit of length. Expressing interstellar distances in miles is like expressing intercontinental distances in millimeters—possible, but awkward. We choose, instead, the light-year, defined to be the distance that light travels in one year. Traveling at a speed of approximately 186,000 miles per second, light covers a distance of nearly 6 trillion miles in one year.

Marking our cosmic yardstick in light-years, we note that the nearest star, Proxima Centauri, is located at a distance of 4.3 light-years from us. Most of the stars that we see in the night sky are tens, hundreds, or thousands of light-years from Earth. The Orion Nebula, visible to us as the central spot of light in Orion's sword, is 1,500 light-years from our terrestrial base.[14]

The Sun, Proxima Centauri, the stars visible to the naked eye, numerous nebulae and stellar clusters, plus hundreds of billions of other stars are arranged in a still larger structure called the Milky Way Galaxy. Our galaxy takes the form of a spiral-armed disk whose diameter spans

14. Cosmic distances are determined by a variety of techniques. The distances to several thousand nearby stars can be measured by a direct geometrical technique (the "annual stellar parallax" method). Based on what has been learned about the properties of these nearer stars, it is possible to compute the distances to many others from the measured values of their apparent brightness (employing the "inverse square law" for light intensity).

a distance of a hundred thousand light-years. To look at the Milky Way visible in the nighttime sky is to look in the plane of our galaxy, the plane in which we see the greatest concentration of neighboring stars, nebulae, and other residents of this ponderous pinwheel.

Impressive as it is, the Milky Way Galaxy is but one of many galaxies. From Earth's southern hemisphere we can see two nearby dwarf galaxies, called the Large Magellanic Cloud and Small Magellanic Cloud respectively. Their distance from the center of the Milky Way is roughly 170,000 light-years. Our nearest neighboring major galaxy, a large spiral much like the Milky Way, is the Andromeda Galaxy, so named for the constellation of nearby Milky Way stars visible in the foreground. The Andromeda Galaxy, sometimes referred to as M 31, is located at a distance of two million light-years.

Beyond the Andromeda Galaxy, in every direction that we look, are billions upon billions of similar galaxies. To the limits of detectability with today's telescopes, we observe galaxies and related objects out to distances on the order of ten billion light-years. At the greatest observable distances, however, the population begins to differ in character from the occupants of the local domain. The relative number of unusual and highly energetic galaxies, such as radio galaxies and Seyfert galaxies, appears to increase with distance. The extreme representatives of this category of peculiar objects are known as quasars. Most quasars are starlike in their photographic appearance; however, from a combination of properties exhibited by their radiation, they have come to be recognized as highly luminous, galaxy-like energy sources. Some quasars emit radiant energy at rates as high as a thousand times the luminosity of an entire normal galaxy; yet the size of their most luminous region may be only a few times larger than the diameter of Pluto's orbit—minuscule by galactic standards! The structure and behavior of these enigmatic luminaries, as well as their place in cosmic history, is the subject of continuing investigation.

The Lookback-Time Phenomenon

First lightning strikes; then we hear the thunder. The delay between a sound-making event and our hearing of the sound is familar to all of us. It arises as a direct consequence of the finite speed of sound. Sound travels at a speed of about a thousand feet per second; in five seconds it travels a mile. When lightning strikes a mile away from us, a sudden burst of sound is generated by that strike. But before we can hear it, that

M 31, the Andromeda Galaxy, the nearest major spiral galaxy to the Milky Way Galaxy, approximately two million light-years distant. Palomar Observatory Photograph

sound must travel the mile from the lightning strike to our ears. To travel that mile takes five seconds; thus we hear the thunder five seconds after the lightning strike occurred. Because sound travels at a finite speed, there is an unavoidable delay between any sound-generating event and our hearing of it. The amount of delay is directly proportional to the distance separating the locations of event and observer.

But light also travels at a finite speed. It takes time for light to travel from its source to an observer. In the course of our ordinary daily activities, the consequent time delay between an event and our reception of light generated by that event is negligibly small. However, when a light source and an observer are separated by astronomical distances, that time delay between event and observation becomes strikingly significant. The Sun, for example, is 93 million miles from Earth; traveling at 186,000 miles per second, light requires approximately eight minutes (500 seconds, to be more precise) to travel from the Sun to planet Earth. The light that we receive from the Sun at any instant is light that left its surface eight minutes earlier. We never see the Sun as it *is,* but only as it *was* eight minutes earlier. If, for instance, we were to observe a solar flare suddenly erupt from the Sun's surface, we would be seeing an event that *had occurred* eight minutes earlier. We call that eight minutes the "lookback time" for the Sun.

Looking out at distant objects, we are unavoidably also looking back in time. Looking at the Sun, we see the Sun not as it *is,* but as it *was,* doing what it was doing eight minutes earlier. Looking at the stars visible to the unaided eye, we see these stars not as they are now, but as they were tens, hundreds, or even thousands of years ago. Their lookback time in years is numerically equal to their distance expressed in light-years. Therefore, when we observe the Orion Nebula, located 1,500 light-years from us, we are seeing those processes and events that were occurring there 1,500 years ago, phenomena that were taking place during the lifetime of St. Augustine.

On February 24, 1987, a supernova was observed to have occurred in the Milky Way's dwarf companion galaxy, the Large Magellanic Cloud. But when did this stellar explosion actually happen? Because the Large Magellanic Cloud is 170,000 light-years distant from earth, its lookback time is 170,000 years. Thus, to observe supernova 1987A was to observe a burst of light that was sent on its way to Earth 170,000 years ago—long before the birth of recorded human history.

The consequences of the lookback-time phenomenon are even more striking when we consider our observation of distant galaxies. The lookback time for the Andromeda Galaxy is two million years. Thus, if we were to observe a supernova event in M 31, we would be observing an event that took place two million years ago. And looking at galaxies in a group called the Virgo cluster, we are seeing light that was generated by processes that were occurring about fifty million years ago. Similarly, the radiation now being received from the radio galaxy Cygnus A was emitted from that source a billion years ago. From the quasar 3C 48 we

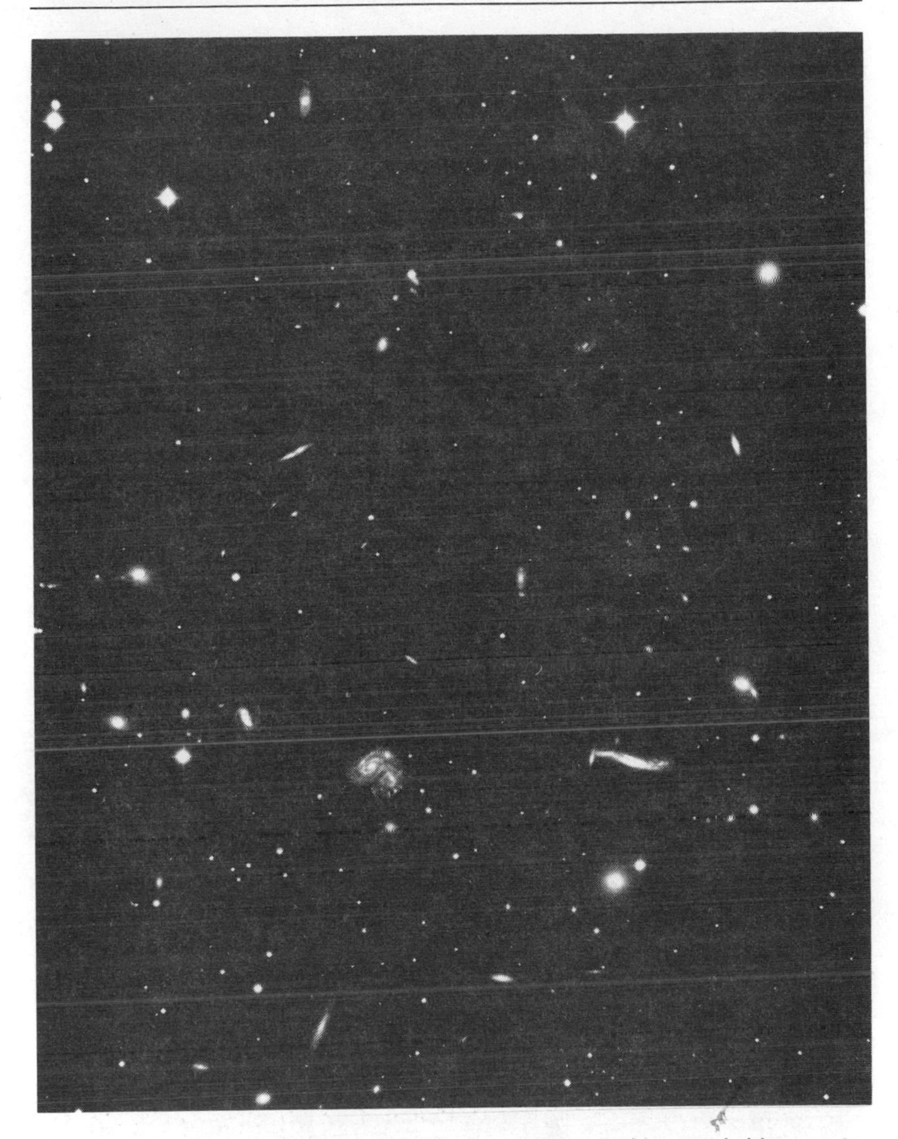

The Hercules Cluster of galaxies. The light from objects in this remarkable grouping has been traveling to Earth for hundreds of millions of years. Palomar Observatory Photograph

are receiving light waves and radio waves that were generated in processes and events that took place five billion years before the present era.

Because of the lookback-time phenomenon, a direct and unavoidable consequence of the finite speed of light, we are able to see cosmic history as it was happening. Cosmic history is on display for direct observation. To learn of this history we need not apply a lengthy chain of reasoning to a myriad of subtle clues; all we need do is look. The light we are receiving is the direct conveyor of information about events and processes that constitute the history of the cosmos on the galactic scale. To look deep into space is to look far back in time. The book of cosmic history is open, waiting to be read.

Quasars as Youthful Galaxies

Do galaxies, like stars, evolve? Were galaxies significantly different in the past?

The study of galactic history is far more difficult than the investigation of stellar history. Compared to stars, galaxies are far more vulnerable to external influences. The tidal distortion of one galaxy by another and collisions between galaxies are examples of phenomena that can make galactic history a complicated matter. Furthermore, because galaxies are structurally far more complex than stars, even their internal evolution (change as a consequence of physical processes) is an extremely difficult phenomenon to model adequately.

There is some evidence, however, that galaxies have experienced episodes or stages in their temporal development that differ markedly from the "normal" state of a galaxy. That evidence is available as a consequence of the lookback-time phenomenon. Looking far back into the past (that is, looking at extremely distant objects), we note an increase in the population of "active galaxies" and related objects. These galaxy-like entities are different from normal galaxies both structurally and energetically. Structurally, the most luminous portion of several classes of these objects is small compared with normal galactic core regions. Energetically, the luminosity values for these peculiar objects can be very large compared to ordinary galactic luminosities. Quasars appear to be extreme in both categories: an extraordinarily high luminosity generated by a relatively small region.

Looking back into cosmic history, we find the population of quasars first increasing, then diminishing. Astronomers make no claim to know exactly what causes these population changes, but it appears very likely that quasars are actually galaxies, seen as they were billions of years ago—that is, in their youth—at which time they were passing through transient phases of prodigious energy generation. As galaxies mature, the frequency and intensity of these energetic outbursts apparently diminishes. Older galaxies, such as the ones seen in our own vicinity, are relatively quiescent. Distant galaxies, because of their proportionately greater lookback time, are seen as they were during their volatile "adolescent" period. Thus, the fact that the population of active galaxies and quasars varies with distance appears to be a direct consequence of galactic evolution. Galaxies were different in the remote past than at present. Much remains, however, to be learned about the temporal development of galaxies.

The Hubble Law

Light from a remote galaxy, like light from a nearby star, is the carrier of information concerning its source. A photograph formed by that light reveals the structure of a galaxy, perhaps a spiral like the Milky Way. Using an instrument called a spectrograph, astronomers are able to determine the wavelength spectrum (the color distribution) of galactic light. An early twentieth-century investigation of galactic spectra, begun around 1914 by Vesto Slipher, revealed that the spectra of most galaxies are shifted toward the red, that is, toward larger wavelength values.[15]

It was soon discovered, however, that there was a significant pattern in the measured values of redshift. During the 1920s, Edwin Hubble made careful determinations of the distances to several galaxies, and discovered a correlation between redshift and distance.

The simplest available interpretation of the observed redshift was that it was caused by the Doppler effect—a shift in the measured wavelength values of light emitted by a source moving along the line of sight.[16] The light from an approaching source would be blueshifted toward smaller wavelength values; a redshift would indicate that the light source is receding. The amount of Doppler redshift is in direct

15. The amount of redshift can be determined by comparing galactic spectra with the spectra of standard lamps; the wavelength distribution of each spectrum is marked by recognizable patterns imposed by the chemical elements. The "redshift parameter," z, is defined to be the ratio of the change in wavelength, $\Delta\lambda$, to the original wavelength value, λ_o.

16. The Doppler effect for sound is perhaps more familiar to us than is the effect for light. The sound from an approaching source appears to have higher pitch (higher frequency, shorter wavelength) than when the same source is at rest. A receding source, on the other hand, appears to emit sound at a lower pitch (lower frequency, longer wavelength). Listen carefully, for example, to the sound of a train's horn or of a propeller-driven airplane as it first approaches, then passes and recedes; the drop in pitch from higher to lower should be quite noticeable.

proportion to the velocity with which the light source is receding from the observer.

Interpreting the galactic redshift as an indicator of recessional velocity, Hubble noted that his data revealed a relationship of direct proportionality between galactic recessional velocity and distance, a relationship that now bears his name—the "Hubble Law." (It is conventionally written as v = Hd; where v is the recessional velocity of a galaxy, d is its distance, and H is the constant of proportionality, now called the "Hubble constant.") In 1929, Hubble presented his initial results to the National Academy of Sciences in a paper titled "A Relation between Distance and Radial Velocity among Extra-Galactic Nebulae." During the next several years, he and Milton Humason continued to gather data and found the Hubble law to be convincingly confirmed. Modern measurements continue to support the thesis that galactic redshift and distance are directly proportional, with the possible exception that at very large distances the redshifts are slightly higher than might be expected from a strict proportionality.

The Expanding Universe Interpretation

If galactic redshift is the consequence of the Doppler effect, and if the Hubble law faithfully reflects the relationship between the recessional velocity and distance of galaxies, then a remarkable conclusion can be drawn: the universe is expanding. Everywhere throughout the cosmos, galaxies are drifting apart from one another. It is as if galaxies were attached to a three-dimensional lattice of giant rubber bands and the entire system were being uniformly stretched in all directions. The inhabitants of any one galaxy would observe all other galaxies to be systematically receding in the manner described by the Hubble law.

Cosmological models are theoretical constructs that attempt to incorporate all of the properties and behavior patterns exhibited by the physical universe as a whole into a single coherent system. Contemporary cosmological models have abandoned the Doppler interpretation of galactic redshift and have modified the concept of cosmic expansion. In the earlier Doppler interpretation, the redshift was taken to be a direct consequence of the recessional motion maintained by the source relative to the observer. Galaxies moving through space relative to an observer would exhibit redshifts in proportion to their recessional velocity. In the new interpretation, the cosmological redshift is perceived to be a consequence of the expansion of the space between galaxies that takes place while the light is en route between source and observer. Galaxies are thought to move, not *through* space, but *with* an expanding spatial framework. And as the light they emit travels through this expanding space from source to observer, wavelength values are stretched to the same degree as the spatial framework is expanded (something like the lattice of giant rubber bands suggested earlier). This stretching of light waves as they travel between galaxies that are comoving with an expanding spatial framework is the cosmological redshift discovered by Hubble. The observed redshift is directly proportional to distance because the amount of stretching that has occurred between the emission of light by a galaxy and the reception of that light by an observer is directly proportional to the lookback time of the galaxy. Light from more distant galaxies must travel for a longer time; therefore it will experience more stretching as the universe steadily expands.

The Age of the Universe

Galactic redshift, interpreted as an indicator of cosmic expansion, implies that galaxies will become progressively farther apart from one another in the future. Projecting backward in time, then, it would seem that galaxies must have been more crowded in the past. In fact, if one were to extrapolate the present behavior of the expanding universe indefinitely far back into the past, one would encounter an intriguing singularity. Going back about 8 billion years, all intergalaxy distances would be about half of their present value—crowded, perhaps, but tolerable. Going twice that far back in time would bring all galaxies to the same place—an intolerable situation! Furthermore, employing the concept of spatial expansion, not only would intergalaxy distances become vanishingly small, but *all* cosmic dimensions would have diminished to zero. Going back to this singular instant in time, the whole universe diminishes to a point.

Our naive extrapolation exercise is fraught with numerous pitfalls (cosmologists are fully aware of these, and proceed with due caution), but even this simplified scenario illustrates some very significant aspects of more realistically constructed models of cosmic history. Of particular interest to us at the moment is the idea that the scientifically accessible history of the physical universe is finite. The universe that we have access to has a finite age that can be computed; its history has a computable beginning. According to contemporary cosmological models, which incorporate evidence drawn from a variety of phenomena, the beginning of the universe took place about fifteen billion years ago, the exact figure depending on the evaluation of certain model parameters. We call this fifteen billion years the "age" of the universe, and we call the first episode of cosmic history the "big bang."

BIG-BANG COSMOLOGY

The Restricted Focus of Scientific Cosmology

Within the domain of natural science (which we discuss further in chap. 5), scientific cosmology is concerned with the structure and history of the cosmos as a whole, not with the peculiarities exhibited by any individual celestial object. In some discussions the term *cosmology* is further restricted to the geometrical study of the character and history of the *space-time structure* of the cosmos, leaving the astrophysical study of the temporal development of the *physical forms* exhibited by its contents to be designated as "cosmogony." For the purposes of our discussion, however, we shall not make that separation; we shall consider cosmogony to be a branch of cosmology. The diversity of physical systems and phenomena that are incorporated into contemporary cosmological theories will be illustrated in the course of our presentation.

The Presuppositions of Scientific Cosmology

The goal of scientific cosmology is to construct a comprehensive cosmological model—a theoretical framework for cosmic history that seeks to account for the present structure of the physical universe and its contents in terms of a coherently patterned succession of physical processes and events. Although some of the subject matter of modern scientific cosmology may seem far removed from the more mundane aspects of natural science, the methodology it employs and the presuppositions on which it is based are essentially the same as for the whole enterprise of professional natural science. Among the standard working assumptions of scientific cosmology are these:

1. The behavior of matter and material systems is governed according to discernible patterns. (Note: since many of these patterns denote only the probabilities for each of several options, this assumption does *not* imply a rigid determinism.)

2. These patterns of material behavior are related to the physical properties of material systems and their environment.
3. These patterns are intelligible and can be discovered by analysis of systematic observation.
4. These patterns are the same everywhere in the universe and at all times throughout its history.
5. These patterns are coherently interrelated.
6. Cosmic history (at least the part of it that is accessible to scientific cosmology) is composed of a continuous flow of processes and events that conform to the patterns described above.
7. Cosmic history is authentic history: the physical universe is not permeated with the superficial appearances of events that never actually took place.

The "Standard Model"

Extrapolating back in time from the present moment, and taking all known physical processes into consideration (if we failed to do this, the result of the extrapolation would be utterly worthless), cosmologists conclude that the "initial state," or first describable state, of the universe must have been characterized by nearly infinite temperature and density. From the first tick of the cosmic clock, the expansion of the universe (including space itself) proceeds. Because of the rapidity of the initial era of cosmic expansion, the primeval episode of cosmic history has been dubbed the "big bang." In the next section we shall make some comments on a misleading connotation suggested by this explosion metaphor, but for the time being we shall use the term in the customary way.

UNTIL A DECADE AGO, most cosmological theorizing was concerned with the large-scale and long-term features of the universe. During the last decade, however, the character of cosmological study has been greatly enriched by the development of theoretical models for the microscopic phenomena that may have occurred during the very first moments of cosmic history—the era during which the elementary constituents of the material world were formed in the unimaginably hot and crowded environment of the "primeval fireball." Contemporary cosmological models are the healthy offspring of the marriage of traditional relativistic models for the structure of space-time with recently developed theories for the physics of elementary particles.[17] Contemporary models are able to incorporate both the large and the small, both the quick and the slow, both the dense and the vacuous, both the hot and the cold. These models are more comprehensive than ever before, coherently integrating more cosmic properties and behavior patterns than most people ever imagined possible.

Following is a very brief description of some of the principal elements in what cosmologists now call the "standard model." Many technical details remain to be worked out, particularly for the first 0.01 second, but the broad outline seems rather securely in place.[18]

THE FIRST FEW SECONDS after the beginning, according to contemporary theory, are characterized by a flurry of important phenomena. The rapid expansion of space is accompanied by a correspondingly rapid drop in temperature. As the temperature falls, the four basic forces (gravity, electromagnetism, and the "strong" and "weak" nuclear forces) make their successive appearances as distinguishable entities, thereby revealing the fundamental patterns of material behavior for all future times. The drop in temperature also permits the fundamental building blocks of matter to exist as relatively stable entities, first the heavier particles (collectively called *hadrons*), then the lighter particles (collectively called *leptons*). A residue of particulate matter is left by this process, the principal form being the nuclei of hydrogen atoms. By the time the universe is a few seconds old, its supply of hydrogen, the fuel required for future stellar processes, has been generated as a product of a rapid succession of processes similar to those that can be studied in the laboratories of high-energy physics.

During the next several minutes, the conditions for the thermonu-

17. By "relativistic models" I mean mathematical descriptions, employing Einstein's general relativity theory, of the geometrical properties of spacetime and its interaction with matter on the cosmic scale.

18. There are several books that provide the nonspecialist with very readable discussions of contemporary cosmological theories. The general astronomy texts listed in note 3, for example, would be adequate for most readers. Timothy Ferris's *The Red Limit*, 2nd ed. (New York: Quill, 1983), is an extraordinarily well-written history of the development of twentieth-century cosmological concepts. Those wishing to go beyond the brief discussions of physical cosmology found in general textbooks are advised to look at Edward R. Harrison's *Cosmology: The Science of the Universe* (Cambridge: Cambridge University Press, 1981). Though this book was written for the nonspecialist, I have found it to be one of the most informative and well-written books on the topic.

clear fusion of hydrogen into helium are satisfied, and about one-fourth of the hydrogen is converted into helium. This mixture of hydrogen and helium is the raw material of which stars will be made at a later time.

Following these first moments, storming with action, there comes an era of relative calm. For a period of approximately 700,000 years, the temperature continues to decline as spatial expansion advances. By the end of this epoch the temperature drops to a value of about 3,000K, low enough to allow the combination of electrons with the nuclei of hydrogen and helium to form electrically neutral atoms for the first time.

This transformation from a mixture of free electrons and nuclei (physicists call it the "plasma" state) to electrically neutral atoms is accompanied by a drastic change in the properties of the cosmic medium. Prior to this event, the plasma was opaque to radiation, forcing the material and radiative components of the universe to remain in thermal equilibrium with one another and to maintain a very smooth spatial distribution. Once the plasma is transformed into neutral atoms, the electromagnetic radiation is free to travel relatively unhindered. Employing the standard terminology, we say that the matter and radiation components of the cosmic medium have become "decoupled"—that is, each component becomes free to act in relative independence from the other. The 3K cosmic background radiation first detected in 1965 by Penzias and Wilson has been identified as the cooled-off remnant of the radiation released at this decoupling event.

As a consequence of continuing cosmic expansion, the conditions conducive to galaxy formation (for example, a certain degree of nonuniformity in the distribution of matter) appear to have been established. Many details concerning the conditions and processes that combined to give birth to galaxies are not yet well understood, but circumstantial evidence strongly suggests that most galaxies were formed during the first few billion years of cosmic history, with quasars representing a highly energetic, transient phase of very young galaxies.

After galaxies become viable structures, the formation of individual stars may take place. The synthesis of helium and heavier atomic nuclei resumes, now in the core regions of stars. The more massive stars mature quickly, ejecting some of the products of stellar nucleosynthesis (the formation of heavier elements by the process of thermonuclear fusion) back into the interstellar medium.

The formation of planetary systems is judged to be a common by-product of stellar birth. Later generations of stars, formed from nebulae whose heavy element content has been enriched by the debris of massive stellar explosions, may have some planets whose chemical composition is dominated by these heavier elements. Planet Earth and the other terrestrial planets fall into this special category.

From here on, the history that concerns us most is the history of Earth's surface and of the living creatures that have inhabited the rich variety of terrestrial environments. Some of that history is discussed by Davis Young in chapter 3. The remainder must be sought elsewhere. Our brief outline of contemporary models of cosmic history ends here.

The Big-Bang Metaphor

According to the standard model, the beginning of cosmic history was characterized by the awesomely rapid expansion of an intensely hot plasma of matter and radiation. Within human experience, the only analogous phenomenon to which this singular event might be compared is an enormous explosion—a hydrogen bomb explosion, for example. But the explosion analogy and the *big-bang* metaphor, however helpful they may be in our attempts to describe something so far beyond our experience, have a misleading connotation.

References to explosions and other things that go bang bring to mind scenes of destruction, devastation, and disorder. When the initial episode of cosmic history is referred to as the "big bang," we quite naturally conjure up the image of a bomb-like event that shatters existing structures and obliterates order. Explosions are *de*structive, not *con*structive.

But the big bang of which cosmologists speak is not at all like a destructive or disordering explosion. Quite the contrary; the initial rapid expansion of the universe is a constructive process that sets the stage and prepares the players for the historical drama that is to follow. Perhaps this primordial episode should be renamed. Instead of "big bang" we might call it the "grand opening," or "act one" of the drama of cosmic history. Or, to try a different metaphor, the first episode is not to be compared with the boom of a nuclear bomb in wartime, but with the bloom of a flower blossom in springtime. These metaphors may fail to have the dramatic appeal of a big bang, but they more faithfully convey the idea that coherence, integrity, order, and structure are developed, not destroyed, during the initial episode of cosmic history.

Doubtless, the big-bang metaphor will prevail; but let the reader be aware that any implication of destruction and disorder must be stripped away. For popularizers of scientific cosmology to overlook this pitfall may be no more than carelessness. But for critics of scientific cosmology to argue that the orderly structures present in the universe *could not* possibly form as products of the big bang is grossly misleading. The big-bang scenario in itself neither guarantees nor precludes the formation of galaxies or stars or planets or people. The rhetoric that attempts to preclude the possibility of an evolutionary cosmic history by arguing that the big bang is an inherently destructive or disordering event is the rhetoric of misapplied metaphor. The big-bang metaphor was introduced to convey the idea of rapid development from a singularly hot and dense beginning. To employ that metaphor in such a way that it gives the appearance of precluding the evolutionary development of ordered structures is carelessly to violate the limits of applicability of an otherwise useful metaphor.

A closely related misconception pertains to the implications of the second law of thermodynamics. The universe presently contains a rich diversity of ordered structures. The early universe, on the other hand, prior to the decoupling event, was in a very nearly homogeneous state. One way of stating the second law of thermodynamics is to say that an isolated system tends to move toward the most disordered state available to it. Naively applied, the second law may appear to preclude the possibility that the present ordered universe could have developed from an earlier homogeneous (disordered) state. However, as physicist Steven Frautschi has argued, the inclusion of the effects of both gravity and spatial expansion in a thermodynamic investigation of the observable universe leads to a very different conclusion: the entropy (a measure of disorder) of the universe has increased from its initial value, but not nearly as rapidly as has the maximum value permitted by the expanding spatial framework.[19] The universe has strayed progressively farther from the equilibrium state of uniform density and temperature, but it is this that permitted the development of such nonequilibrium structures as galaxies and stars.

According to Frautschi's analysis, the universe is not moving toward a "heat death" (a state devoid of energy exchange because of uniform temperature) as nineteenth-century theorists envisioned. The universe is neither maintaining nor moving toward a state of thermodynamic equilibrium. On the contrary, because of the effects of gravity in an expanding spatial framework, the universe is moving progressively farther from equilibrium, thereby maintaining an environment in which

19. Frautschi, "Entropy in an Expanding Universe," *Science* 217 (13 August 1982): 593-99.

energy exchange processes can take place over an extended period of time.

Our advice to the reader, therefore, is this: be extremely wary of assertions that the universe, as presently structured, could not have developed from a big-bang beginning. Neither the misapplication of the big-bang metaphor nor the misconstrual of the second law of thermodynamics should be granted the status of adequate argumentation. Christian scholarship is surely not advanced by such strategies.

The Big Bang and Creation ex nihilo

We have spoken of the big bang as the opening episode of cosmic history—the "beginning." But as Christians we quite naturally associate the concept of a beginning with the opening line of Genesis 1—the majestic announcement with which the first creation narrative opens: "In the beginning God created the heavens and the earth." It is tempting, therefore, to equate these two "beginnings," and thus to view the big-bang theory as a scientific version of creation *ex nihilo*.

However, before two things can be equated, or even be made the objects of direct comparison, it must be demonstrated that they belong to the same category. Are the concepts of a big-bang beginning and creation *ex nihilo* members of the same category? Do they speak to the same issues? Do they provide answers to the same questions? We believe that although there may be a historically rooted point of contact, these two concepts differ in substantial ways and should therefore be treated as distinctly different ideas.

As we have seen, the scientific concept of a big-bang cosmological model focuses its primary attention on the early portion of the formative history of the physical universe. Its concern is limited to physical processes and their contribution to the present structure of the universe and its physical components. The ultimate question of cosmogenesis—How is it that there is something rather than nothing?—lies beyond the domain of scientific cosmology. Similarly, other questions of a metaphysical or religious character—questions about the identity of the universe and about its status in relationship to deity, for example—also lie outside of the scientific domain. Tracing the principal lines of cosmic history back to the first instant of time (time itself being a component of the physical cosmos) might be a remarkable achievement for physical cosmology, but it leaves several profound questions to be answered by other means.

And what of the concept of creation *ex nihilo?* What is its subject matter? What questions and concerns does it address?

Historically, the concept of creation *ex nihilo* has often served an important polemical function—namely, to contradict the concepts of the identity and status of the world offered by dualism and by emanationism. In opposition to Greek dualistic thought, the Christian community has long denied that the physical world may be viewed as some autonomous power coequal with God or as some inherently evil entity in contention with God. Drawn or extrapolated from such biblical passages as Genesis 1:1, Hebrews 11:3, and Romans 4:17, the concept of creation *ex nihilo* served to guarantee that the world could not be coequal with God. Similarly, if the world were created by God *from nothing,* then it could never be seen as emanating *from God* himself. Thus, the unacceptable perspectives of cosmic identity and status offered by dualism and by emanationism were simultaneously rejected, and in their place the Christian doctrine of creation *ex nihilo* proclaimed that the cosmos was totally other than God, yet completely dependent on him for both its existence and its governance.

But this doctrinal concept, with its thoroughly theistic answers to several profound questions of a *metaphysical* character, has long been closely associated with the picture of a *historical* act of divine inception at the world's beginning. How closely are these metaphysical and historical features of doctrinal tradition related? Are they inseparable parts of a single package? In *Maker of Heaven and Earth,* an extensive discussion of the historic Christian doctrine of creation, theologian Langdon Gilkey carefully develops the distinction between questions regarding the historical particulars of God's creative work and the ultimate metaphysical and religious questions regarding the origin and meaning of the world's existence. According to Gilkey, the central meaning of the doctrine of *creatio ex nihilo* is threefold: "that God is the transcendent source of all existence; that creaturely existence is dependent, contingent, and transient, and yet possesses a reality and a value in its own fulfillment; and finally, that the divine act of creation is to be understood not in terms of structure but in terms of its divine purpose, as a free act of a loving will."[20] Within the bounds of this theological perspective one could meaningfully consider a diversity of historical scenarios for the manner and timetable that God might have chosen to perform his creative activity.

20. Gilkey, *Maker of Heaven and Earth* (Garden City, NY: Doubleday, 1959), p. 73. Chapters 2, 3, and 9 of this work are especially relevant to the present discussion. See also chapters 7 and 8 of Gilkey's *Creationism on Trial* (Minneapolis: Winston Press, 1985).

One scenario, a picture that matured in the context of the medieval static-universe concept[21] of the cosmos as a set of fixed structures and species (a picture which anachronistically persists in parts of the Christian community even today), envisions God at the beginning of time calling the material of the universe into existence and then, during the ensuing six-day period, impressing on that material the various structures that we see today: galaxies drifting apart from one another as if from a common point; stars generating light as if by thermonuclear fusion induced by gravitational collapse; planets, complete with the superficial appearance of a complex geological history; and an array of living creatures, complete with the appearance of temporal and genetic relationships.

An alternative scenario for cosmic formation—one consonant with contemporary cosmological theory—envisions God calling the Creation into existence in nascent form and directing its formation in the coherent manner described by the standard big-bang model of scientific cosmology.

Taking this into account, we again ask whether we can equate the concepts of big-bang beginning and creation *ex nihilo.* With considerable conviction we maintain that this would be most inappropriate. Creation *ex nihilo,* we have emphasized, is fundamentally a metaphysical concept that offers a biblically based, theistic answer to the question of cosmogenesis: How does the universe come to have existence? The big-bang episode, on the other hand, offers no explanation of cosmic existence but only a description of the early formative development of the cosmos that exists. And while the big-bang scenario offered by the standard cosmological model might conform to one's *picture* of an act of divine inception, it wholly lacks the *theological substance* of an adequate concept of creation *ex nihilo.* The big-bang model, with its idea of a singular and spectacular beginning may serve to reinforce the traditional Christian concept of the finite temporal duration of the created realm, but it should neither be identified with nor counted as compelling evidence for the doctrine of creation *ex nihilo.* Creation *ex nihilo* is a rich theological concept not merely about temporal beginnings but concerning the fundamental identity of the world and the source of its existence at all times; the big-bang model is a theoretical scientific concept limited to the description of selected aspects of the formative history of the physical universe. A big-bang beginning and creation *ex nihilo* cannot be equated. In no way do they offer answers to the same question.

21. For a full development of this concept, see N. Max Wildiers, *The Theologian and His Universe,* trans. Paul Dunphy (New York: Seabury Press, 1982).

IS COSMIC HISTORY EVOLUTIONARY IN CHARACTER?

In our discussion of several aspects of cosmic history, we have often used the term *evolution*. We have spoken of stellar evolution, galactic evolution, and planetary evolution. Is it legitimate to say, then, that the whole of cosmic history is evolutionary in character? To many members of the Christian community such a statement might appear to be at odds with traditional pictures of God's work as the Creator, his calling the cosmos into existence and giving form to its contents. And when an evolutionary scenario for cosmic history is perceived to be inconsistent with traditional *pictures* of divine creative activity, it may also be judged by many to be inconsistent with both the biblical word regarding creation and the historic Christian doctrine of creation.

In the remainder of this chapter we shall briefly assess the appropriateness of employing the term *evolutionary* to describe the character of the formative history of the universe. To that end we shall first reflect on some of the characteristic properties of cosmic history as presently understood, then note what qualities the scientific use of the term *evolutionary* ordinarily incorporates, and finally point out some concepts that the scientific employment of this term does *not* logically entail. Concerning the statement "cosmic history is evolutionary in character," our conclusion will be that, while it may serve well as a descriptive statement within the restricted domain of natural science, it utterly fails to supply the logical warrant for a naturalistic creed.

Cosmic History as We Presently Understand It

From numerous lines of evidence, some of it outlined above, we conclude that *the universe is an arena of dynamic change rather than statically preserved structures.* Stars, for example, are understood to be forming by the gravitational collapse of globules of gas and dust within large interstellar nebulae. Once formed, stars continue to change as a consequence of thermonuclear fusion processes occurring within them. Eventually, gravity crushes giant stars into diminished stellar remnants such as white dwarfs, neutron stars, and black holes. The systematic motion of galaxies reveals that the universe is expanding. The standard big-bang model is built upon an extrapolation of this behavior back in time. Planets, including the Earth, are the sites of a rich spectrum of geological phenomena whose history has quite literally given shape to the surface of our terrestrial home. It has become increasingly clear that the contents of the physical universe have been brought to their presently manifested forms by a formative process.

This concept of *dynamically changing* structures stands in direct contrast to any concept of the cosmos as composed of a set of *statically fixed* structures. The concept of the created world as a statically ordered system, however, is a common element in traditional Christian thought. Deeply rooted in the medieval world picture, this concept was formed from a synthesis of Greek science, biblical interpretation, and Christian doctrine. N. Max Wildiers summarizes the world picture of medieval theology as follows:

> For the medieval theologian, then, the world was a perfectly ordered whole. . . . The world was also defined as an ordered collection of creatures. . . . There was no doubt that this order is immutable: it dates from the creation of the world. . . . Divine wisdom has once and for all clearly distinguished one thing from another, assigning everything its proper place in the whole. It is by looking at the stars that we have a foretaste of the imperishableness of God's creation and of the permanence of his world order. What is clear from such a picture is that there is no question of there being a gradual construction of order in the course of history.[22]

Many of us who have grown up within the Christian community will recognize this picture of the created world as a part of our own cultural inheritance.

Quite understandably, medieval theologians sought to interpret biblical references to the physical world in terms of the world pictures of their day. The concepts of cosmic structure and history employed in theological discourse were concepts appropriate to the state of knowledge *at that time.* They included the ideas that cosmic structure was geocentric and that the forms exhibited by the contents of the cosmos were unchangeable.

But now, late in the twentieth century, we are convinced by the results of scientific investigation that the universe is not geocentric in its structure. Consequently, and very wisely, we no longer demand that the Bible be interpreted as if it taught that the universe had been constructed in accord with Ptolemy's Earth-centered model. On the matter of cosmic history, however, a large segment of the Christian community continues to cling to the medieval concept of the immutability of forms. In spite of

22. Wildiers, *The Theologian and His Universe,* p. 57.

an abundance of evidence to the contrary, many Christians persist in holding to a static concept of cosmic history.

Taking our clues from the present behavior of physical systems and the physical record generated by similar behavior in the past, we also conclude that *the processes and events that constitute cosmic history have been governed according to certain intelligible patterns.* Furthermore, the evidence very strongly suggests (1) that these patterns are coherently interrelated, (2) that they are consistent with the concept of proximate causality, and (3) that they have been followed continuously throughout the duration of cosmic history.

Coherence is one of the outstanding properties of the scenario for cosmic history proposed by scientific cosmology. The standard cosmological model, of which we have given only partial sketches, is not merely a list of isolated events and processes that are thought to have occurred at various times and places. Rather, cosmic history appears to be composed of a succession of universal phenomena that are coherently interrelated. Why is this the case? First, because all of the physical processes and events conform to the same patterns for material behavior—patterns that evidence indicates are spatially and temporally invariant. Second, because the outcome of any one event or process establishes the character of both the physical system and the environment for the next event. Each successive state of the physical universe is intimately related to the previous state, and the phenomena experienced by one part of the cosmos are coherently related to phenomena experienced by other parts.[23] Cosmic history seems better compared with a coherently patterned tapestry than with a random patchwork quilt.

The second concept that appears on our brief list of qualities exhibited by physical phenomena is *causality.* To say that the processes that constitute cosmic history are consistent with the concept of proximate causality is to say that they are ordinary physical phenomena, the kind that occur daily. When, for example, we see an apple fall from a tree, we might say that its fall was caused by the force of gravity. We are thereby identifying the force of gravity as the proximate cause for the earthward acceleration of the apple. We say "proximate" because we

23. As we noted in our list of the presuppositions of scientific cosmology, our recognition of the causal connections of physical events may not be interpreted to imply a rigid determinism. The behavior displayed by physical systems appears to be neither unregulated nor completely determined but rather patterned in such a way as to permit a degree of novelty and contingency that is functionally important.

have not explained the ultimate source of gravitational force; we have only described a correlation between the mass and acceleration values of the Earth and the apple. Furthermore, though we commonly call this a "cause-effect relationship," it may be more appropriate to call it a "properties-behavior relationship." Regardless of which terminology is preferred, we do wish to recognize that the construction of cosmological models from phenomena that are wholly consistent with the concept of proximate causality has been a most fruitful enterprise.

Finally, to say that the ordinary patterns for material behavior have been followed *continuously* throughout cosmic history is to exclude from cosmological models the introduction of arbitrary discontinuities. It is becoming increasingly evident that the interpretation of the physical record of cosmic and terrestrial history does not require a reliance on any special events that interrupt or contravene the ordinary patterns of proximate causality. (The obvious exception to this is the inception of the universe at the beginning of time, but this singular event of cosmogenesis appears to be indescribable by any cosmological model.) That part of cosmic history accessible to scientific investigation appears to be composed of a continuous flow of phenomena that conform to the ordinary patterns for the behavior of physical systems.

Some people might raise questions concerning the authenticity of the coherence, causality, and continuity that our scientifically derived scenario for cosmic history displays. Are these qualities legitimately ascribed to the actual history of the universe, or are they merely artifacts of the presuppositions on which modern scientific cosmology is based?

We believe that the coherence, causality, and continuity to which we have drawn attention are not merely the contrived properties of the standard cosmological model but are indicative of the character of actual cosmic history. Though our brief presentation may fail to demonstrate this adequately, the evidence supporting the concept of a coherent and continuous cosmic history along the lines of the scenario described earlier is to us very convincing. This evidence has been drawn from a wide diversity of phenomena within the domains of geology, astrophysics, physical cosmology, and other natural sciences. We judge that the community of natural scientists is investigating this history with appropriate methodology and with remarkable competence. We judge that the results of their investigations are being reported with admirable integrity in the professional literature. Our conclusion, therefore, is that these results must be taken seriously. They cannot be dismissed lightly. Any proposal for an alternative scenario for cosmic history must account for a vast amount of empirical data and must offer an equally adequate means of integrating those data into a coherent explanatory framework. Appeals to occasional anomalies or unsolved puzzles will not suffice as an adequate basis for dismissing the standard model.

The Christian community, we believe, gains nothing by closing its eyes to the results of scientific cosmology. One might even argue that an adequate Christian witness to a scientifically well-informed culture *demands* a thoughtful employment of these results.

Having drawn attention to cosmic history as exhibiting dynamic change according to intelligible patterns, we now also note that *these changes are cumulative and give cosmic history a directional character.* Consider the following: planets, such as our home planet Earth, appear to be the by-products of stellar birth—formed from material left in orbit around a newborn Sun. But the heavy elements that are such an important component in the chemical makeup of terrestrial planets and their living inhabitants are understood to be the products of thermonuclear fusion processes and supernova events that occurred earlier in the history of the Milky Way Galaxy. And the large interstellar nebulae that provide the material and the physical environment conducive to the formation of stars like the Sun require the context provided by the entire spiral-structured galaxy that we inhabit. But even galaxy formation is seen as an integral part of cosmic history. Episodes of galaxy formation could not begin to take place until the radiation and material components of the cosmic medium were decoupled, allowing gravity to generate a "lumpiness" in the spatial distribution of hydrogen and helium. Even the hydrogen and helium that make up the majority of solar material are seen as the products of specific processes that took place in the first few minutes of cosmic history. And the whole succession of phenomena that we have just listed appears to require the context provided by an expanding spatial framework—an expanding universe of space-time, matter, and radiation interacting with one another within the bounds of intelligible patterns of physical behavior.

In summary, the evidence to which we have referred, and a vast amount of similar evidence, very strongly suggest that our presence on planet Earth, as creatures with bodies having a particular chemical makeup and requiring a particular physical, chemical, and biological environment, is dependent upon the presence of the entire physical universe and the occurrence of the whole of cosmic history. Cosmic history, therefore, is not merely of academic interest. Rather, if our present understanding of it is correct, cosmic history represents the fifteen billion years of formative preparation that has been directed toward the viability of our presence as creatures living on planet Earth.

Cosmic Evolution as a Scientific Concept

The standard cosmological model is generally referred to as an "evolutionary" model. For many members of the Christian community, the association of this label with a product of scientific theorizing is taken to be sufficient warrant for a categorical rejection of large segments of the professional scientific enterprise. But what does this label "evolutionary" actually entail? In the context of scientific cosmology, it entails essentially the same qualities for cosmic history that we have just been discussing—cumulative, directional development as a consequence of the behavior of physical systems according to patterns that are coherently interrelated, exhibit proximate causality, and are continuously applicable for the duration of cosmic history.

As a concept restricted to the domain of natural science (to be discussed further in chap. 5), cosmic evolution is concerned only with the formative history of the physical universe. As a scientific concept, it is capable of dealing only with matters regarding the intrinsic intelligibility of cosmic history. The standard model for cosmic evolution offers a set of answers to questions concerning what events and processes may have taken place—questions about the path traced out by cosmic history and the chronology of that journey. As a scientific concept, evolution is more a matter of partial *description* than of complete *explanation*.

As a descriptive scientific concept, cosmic evolution need not be rejected by the Christian community. While it may call for the abandonment of a medieval *picture* of the cosmos as being an arena for the preservation of immutable forms, it need not require, or even suggest, the abandonment of a biblically faithful theological *doctrine* concerning the cosmos as God's Creation. Having the status of Creation, the cosmos stands in total dependence on God as the source, or origin, of its existence—for its preservation no less than for its inception. As Creation, the physical universe is governed according to patterns established and sustained by the sovereign Creator. As the history of the Creation, cosmic history accomplishes the purposes of its divine Governor. And as Creation, the world and all of its creaturely inhabitants find their value in their relationship to the Creator.

Naturalistic Evolutionism as a Creedal Perspective

Having made the claim that the scientific concept of cosmic evolution is not inherently at odds with Christian belief, we must give full recognition to the fact that there are numerous outspoken members of contemporary Western culture who do, with considerable vigor, assert that this scientific concept provides the logical warrant for their naturalist worldview. *Naturalistic* evolution*ism* is the product of an attempt to employ the concept of cosmic evolution *creedally,* not just *scientifically.*

As Christians, we have a responsibility to speak with boldness and confidence against the naturalistic evolutionism of our day. But we must do so thoughtfully, making use of important distinctions. The scientific concept must be distinguished from the creedal perspective. Evolution must be distinguished from evolution*ism.* The term evolution*ism* should be reserved for the naturalistic creedal position, and the term evolution*ist* should be reserved for those who seek to indenture the scientific concept of evolution in the service of their naturalistic creed. Cosmic evolution is a scientific concept. Evolution*ism* is a naturalistic creedal perspective.

Naturalistic evolution*ism* is to be rejected because its materialist creed puts the material world in place of God, because it asserts that the cosmos is self-existent and self-governing, because it sees no value in anything beyond the material thing itself, because it asserts that cosmic history has no purpose, that the idea of purpose is only an illusion. Naturalistic evolutionism views the cosmos as an independent, autonomous, material machine named NATURE—a singularly meaningless image compared with the rich biblical vision of the cosmos as God's CREATION.

The Christian community must speak in a way that exposes the fallacy of the naturalist creed. Some proponents of naturalism claim that there are logical bridges that lead from the scientific concept of cosmic evolution to the creedal position of naturalism. We must show that such bridges will collapse under the weight of careful scrutiny.

The following section contains examples, all too briefly discussed, of creedal matters that are not logically entailed by the scientific concept of cosmic evolution. Some of them are specific instances of more general considerations that are applicable not only to evolutionary concepts but to all scientific theories (which are discussed at greater length in chap. 5). Employing a series of categories introduced earlier, we shall refer to creedal tenets concerning status, origin, governance, purpose, and value.

What the Scientific Concept of Evolution Does Not Entail

1. *As a scientific concept, cosmic evolution does not entail a particular specification of the status of the universe relative to God.* Questions concerning the

status of the cosmos (i.e., its identity and its relationship to deity) fall outside of the domain of natural science. The scientific concept of cosmic evolution is limited to matters of intrinsic intelligibility and is silent on matters of external relationship to nonphysical beings. The philosophical materialist might assert that the physical world is a self-existent entity that is wholly independent of any deity. But that assertion must be recognized as a profession of atheistic faith; it is not a conclusion that can be drawn from scientific cosmology or from any other natural science. The concept of cosmic evolution does not provide the logical warrant for naturalism, nor does it contest in any way the status of the cosmos as God's Creation.

2. *As a scientific concept, cosmic evolution does not provide an explanation for the origin of the universe.* Much of the material written about contemporary cosmology, particularly material written at the popular level, speaks of modern cosmological models as theories about the *origin* of the cosmos. But the reader must be alert to the diversity of meanings given to that term. When speaking about the origin of the Sun, or the origin of the Earth, or the origin of the Grand Canyon, we are using the term *origin* to represent the processes by which these specific structures were *formed*. To avoid confusion, perhaps the word *formation* should be used in place of *origin*.

When we ask about the origin of the entire universe, however, we are using the term in a different and far more profound way. In that case we are asking not merely about the *formation* of its structures; we are asking the ultimate question of cosmogenesis: What is the source, or cause, for the existence of the whole cosmos? We are not asking *how* things were formed, but *why* there exists something rather than nothing. Like the question of status, the question of origin (in this ontic sense) lies outside the domain of natural science. The scientific concept of cosmic evolution contributes nothing to its answer. Natural science might suggest that the answer is not self-evident, that the existence of the cosmos is not self-explanatory, but beyond that it must stand in respectful silence.

Not all persons, however, are willing to recognize the boundaries of the scientific domain as we have drawn them. In his brief book *The Creation*, P. W. Atkins argues that it may soon be possible scientifically to explain how the universe came into being from "nothing."[24] It is amus-

24. Atkins, *The Creation* (New York: W. H. Freeman, 1981). For a more extended critique of this book, see Howard Van Till, Davis A. Young, and Clarence Menninga, *Science Held Hostage: What's Wrong with Creation Science AND Evolutionism* (Downers Grove, IL: InterVarsity Press, 1988), pp. 141-53.

ing to note, however, that such self-creating universes always begin with something—a "vacuum state," for example, complete with the power for self-governance and with a blueprint for its transformation into a material state. The "nothing" with which the cosmos of the naturalist begins is not the true absence of anything; it is merely a certain nonmaterial state of this cosmos. Even self-creating universes must begin with some form of "self," thereby leaving the ultimate question of origin unanswered. An observation made by Langdon Gilkey seems especially relevant here: "The vast new powers of science do not, in the end, make religious faith and commitment irrelevant; they make them more necessary than ever."[25]

3. *Evolutionary processes are not inherently naturalistic.* By "naturalistic" we mean, in this context, autonomous as opposed to theonomous—self-governed as opposed to God-governed. As we pointed out earlier, the processes that contribute to cosmic evolution are the ordinary physical phenomena of everyday occurrence. To call a process "evolutionary" may direct our attention to its conformity to the regular patterns for material behavior, but in no way does that entail a specification of the agent that *governs* material behavior in accordance with those patterns. To recognize, for example, that the phenomenon of lightning occurs in accordance with all of the patterns for electromagnetic activity (as described by Maxwell's equations) entails no denial that God is the Governor of his Creation, the one who "sends lightning with the rain" (Ps. 135:7). To recognize patterns of behavior is one matter; to specify the identity of the governing agent is an entirely different matter. Questions of behavior and questions of governance are profoundly distinct from one another and must be dealt with separately. Scientific cosmology can deal effectively with certain aspects of cosmic behavior, but it is powerless to answer questions concerning the identity of the governor of that behavior. Cosmic evolution is no more naturalistic than rainfall or sunrise.

4. *Evolutionary processes are not inherently devoid of purpose.* Because of the statistical nature of some patterns of material behavior, and because a system's behavior is contingent on environmental factors, some people may be tempted to assert that the path followed by any particular evolutionary process is the product of chance alone—and thus devoid of purpose. Evolutionary processes, claim the proponents of naturalism, are composed of a string of accidents; surely a string of accidental events could not function to achieve a specified purpose.

25. Gilkey, *Religion and the Scientific Future* (New York: Harper & Row, 1970), p. 98.

On the surface, this may appear to be a compelling scientific argument. It is, however, neither compelling nor within the domain of natural science. Questions of purpose must of necessity entertain the possibility that the universe and its temporal development achieve their significance in their relationship to a nonphysical realm. Plato's theory of ideal forms and Aristotle's concept of teleological causation gave substantial recognition to that dimension. And the Christian doctrine of creation certainly entails the profession that the physical realm and its history function principally to accomplish the purposes of God. Questions of purpose are fundamentally religious questions; they are not questions that natural science alone is capable of addressing. The evaluation of the purpose of cosmic history, whether that history is evolutionary or otherwise, must be carried out in an arena beyond the bounds of natural science. Consequently, any argument against the presence of purpose in cosmic history that claims scientific warrant for its conclusion will fail to be compelling. (Even argumentation favoring the idea of purpose must be limited. Though the perceived directionality of the evolutionary history of the universe may strongly imply that some purposes are at work, there is no way scientifically to identify those purposes or to demonstrate precisely the mechanism by which they are achieved.)

5. *The evolutionary character of the formative processes at work in cosmic history is neither normative for, nor necessarily extendable into, the arena of social, ethical, or religious values.* We shall not develop this point here beyond noting once again that the question of ultimate values falls outside of the domain of natural science. In formulating a system of values, one must draw guidance from resources other than the character of material behavior alone.

IS COSMIC HISTORY evolutionary in character? In working toward an answer to this question we have found it necessary to make an assessment from two distinct standpoints—scientific and creedal.

From the scientific standpoint, cosmic evolution is a matter of the intrinsic intelligibility of the properties, behavior, and history of the physical universe. The scientific adequacy of the standard big-bang cosmological model must be judged on the basis of its ability to account for the relevant empirical data by means of a coherent theoretical framework. This judgment, we believe, is being performed with both competence and integrity by the community of professional natural scientists, many of whom are Christians.

From the creedal standpoint, however, we must be alert to the manner in which the scientific concept of cosmic evolution is being employed. Naturalistic evolutionism attempts to employ the concept to provide warrant for a naturalistic creed. We judge that attempt doomed to failure because it fails to recognize the limited domain of natural science. Naturalism attempts to warrant creedal tenets regarding status, origin, governance, purpose, and value (matters of external relationship) by appealing to the properties, behavior, and history exhibited by the physical universe (matters of inherent intelligibility). How could such an effort be anything more than an exercise in futility?

5. THE CHARACTER OF CONTEMPORARY NATURAL SCIENCE

HOWARD J. VAN TILL

Having already called attention to many of the achievements of scientific investigation, we now reflect on the character of that enterprise. What kinds of questions are the natural sciences competent to investigate? What are the limits or boundaries of the domain of authentic scientific inquiry? Upon what criteria are the quality of scientific work and the merits of scientific theories judged? How might the worldview commitments of a scientist affect these criteria?

After discussing such questions in a very general way in this chapter, Howard Van Till draws from the science of astronomy a specific case study in which the characteristic features of contemporary scientific investigation are amply illustrated. The intriguing peculiarities exhibited by star clusters provide the occasion for a classic example of scientific puzzle solving.

IN THE PREVIOUS TWO CHAPTERS we presented some of the results of modern scientific investigations into the formative history of the physical universe. In the course of presenting these results we occasionally paused to reflect on the character of the scientific enterprise in general (recall, for example, our listing of the presuppositions of physical cosmology), but we offered no extensive critique. Now we wish to examine more closely the character and limits of natural science in order to assess the credibility and significance of its findings. In chapter 7 we will assess the place and import of these scientific reconstructions from a biblical perspective.

In this chapter we will deal with two major questions and then consider an illustrative case study. First, What does contemporary natural science investigate? We will work toward an answer to this question by discussing first the *object* of investigation and then the *domain* of inquiry concerning that object. Second, What is the operative value system within the scientific enterprise? Our principal concern will be to identify the "epistemic" values that function in scientific theorizing—that is, the system of criteria for evaluating a scientific theory for its likelihood of being a faithful or adequate representation of the actual state of affairs. Finally, as a means of illustrating some of these marks of the contemporary scientific enterprise we will consider the case of star clusters—groups of stars that are gravitationally clustered in a relatively compact arrangement and that exhibit an intriguingly peculiar set of properties.

WHAT DO THE NATURAL SCIENCES INVESTIGATE?

The Object of Investigation

Perhaps before we proceed any further we should remind ourselves that by the term *natural sciences* we simply mean those endeavors known by such names as physics, chemistry, geology, astronomy, and biology, along with their subdivisions and numerous related or combined disciplines. The natural sciences are closely interrelated; in fact, they often overlap and shift their boundaries as specialized sciences develop. You could get a sense of the present location of the boundaries among the several sciences by perusing the subject matter treated in the professional scientific journals, but we are not concerned at the moment with these subdivisions; we prefer to speak collectively of the entire family of natural sciences. Quite intentionally, we exclude from this discussion the various disciplines that are concerned with *human* behavior. Human beings, we believe, cannot be comprehended solely in terms of the physical phenomena accessible to the natural sciences.

What does this family of disciplines study? What is the *object* that is investigated by the natural sciences? To these questions we give a very direct answer: the object of study by the natural sciences is the *physical universe*—no more, no less. It is the world of atoms and of the subatomic particles constituting them. It is the world of things made of atoms:

molecules, rocks, stars, galaxies, living cells, plants, and animals. Whatever can fruitfully be viewed as a physical system—a set of mutually interacting physical entities—can be the object of investigation by the natural sciences.

Nonphysical things are not the object of study by these sciences. The natural sciences in no way deny the existence of other realms or aspects of reality, but their attention is confined to physical entities alone. Consistent with this focus of attention is a generally recognized restriction in the method of acquiring information. Modern natural science is inextricably associated with the empirical (i.e., observational) approach. The object of scientific investigation must be empirically accessible: there must be some way to interact with it physically. Anything that is empirically inaccessible, that provides no avenue of physical interaction, cannot function as the object of study by the natural sciences. The nonphysical realms—important realms indeed—must be investigated by other means.

It should be pointed out that these restrictions on the object and methodology of investigation are not the product of arbitrary choice, nor are they externally imposed. Rather, they are the result of the experience and critical reflection of the professional scientific community. *They are time-tested choices that have demonstrated their fruitfulness.*

The Domain of Inquiry

Although the entire physical universe may be the *object* of investigation by the natural sciences, not all of its qualities and/or aspects fall within the *domain* of such scientific inquiry. But before we approach this matter formally, let us illustrate the distinction between object and domain (as we use these terms) with an example that is close at hand.

The words that appear on this page are formed by a particular distribution of ink on paper. By employing all of the tools of natural science we may be able to determine the identity and, within certain limits, the location of every atom constituting this printed page. But this scientific description of the variety and spatial distribution of atoms would not provide a full account of the object under scrutiny. No matter how complete our description from the standpoint of natural science, it would still fail to reveal an important and very real property of this page—the fact that this particular distribution of atoms and molecules forms words and that these words convey a certain message. Natural science is an appropriate and powerful tool for investigating and gaining knowledge about the physical features of the object of its study, but it is wholly incapable of discovering its *meaning.* To say that this page is *nothing but* a particular assembly of atoms and molecules or to assert that the physical universe is "all there is or ever was or ever will be"[1] is to speak nonsense. In his book *The Clockwork Image,* Donald MacKay calls this nonsense by the colorful but appropriate name "nothing buttery." Furthermore, to assert that natural science is competent to answer all meaningful questions about reality or that only those questions answerable by natural science are truly meaningful is also nonsense. Such claims are not entailed by natural science itself; rather, they are the philosophical/religious assertions of a perspective known as *scientism.* To have confidence that careful empirical investigation will lead to much useful and authentic knowledge about the physical world is an attitude appropriate for the natural sciences. But to assert that these sciences provide the *only pathway* to knowledge about the *whole* of reality is to proclaim the heresy of *scientism.* One can find numerous examples of scientists who make such claims, but when they do so, they are speaking as individuals, not as the voice of the scientific enterprise.

Speaking more formally now, to identify the *domain* of natural science is to identify the categories of questions that it is competent to address. As our example of the printed page has illustrated, natural science is not capable of dealing with all conceivable questions about a physical object; only certain categories of questions lie within its domain. We shall approach the matter of identifying the boundary of this domain by citing several additional examples of questions that clearly lie within the scientific domain and then noting the categories into which these questions naturally fit. We will, of course, keep in mind that these must all be questions about the proper *object* of scientific investigation—the physical universe and its constituent parts.

What is the surface temperature of the star Betelgeuse? What is the quantitative value of the proton mass? What is the structure of a DNA molecule? All of these are appropriate questions for scientific investigation. Each of them fits into the category of questions concerning the *physical properties* of physical objects. All questions that fit into this category lie within the domain of natural science.

Consider another family of questions: What physical process is responsible for maintaining the surface temperature of Betelgeuse? What takes place when an acid and a base are combined? What physical

1. The words are those of Carl Sagan, *Cosmos* (New York: Random House, 1980), p. 4.

changes occur in the process of photosynthesis? These, too, are questions that the natural sciences are capable of investigating. Each of these questions fits squarely within the general category of questions concerning the *physical behavior* of some physical system, and, as such, it lies within the domain of natural science. As a matter of fact, one of the major endeavors in natural science is to construct adequate and accurate descriptions of such phenomena and to discover the universally applicable patterns of physical behavior exhibited by systems with like properties.

Finally, a third family of related questions: What sequence of physical events and processes has contributed to the formation of the Grand Canyon? What occurred on the surface of the Moon to form the craters and other features visible to us? What is the history of life forms on Earth? Does the visible universe of dispersing galaxies have a discoverable history? If so, what is the character and chronology of that history? While these questions may seem somewhat more difficult than most of those cited earlier, these too are examples of questions that are open to investigation by the natural sciences.[2] Questions like these fall into the broad category of questions concerning the *formative history* of the Earth and its inhabitants, of other bodies in the solar system, and of the entire observable universe. On the basis of what has been discovered during the past century or two, we judge that these fascinating questions concerning the formative history of physical systems, including living systems, also lie within the domain of natural science and are fruitful questions that merit careful scientific investigation. Obviously this has been a working assumption throughout our discussions in the previous chapters.

At the risk of oversimplification, we shall say that all (or certainly the vast majority) of the questions that lie within the domain of natural science can be fittingly placed within one of the three categories just introduced: the categories of *physical properties, physical behavior,* and *formative history.* Furthermore, we find it helpful to note that these three aspects of the physical universe are empirically accessible to us because of its *intrinsic intelligibility.* But that very formal term requires further clarification.

By observation, experience, and reflection, the human race has discovered that the physical world is intelligible—that is, it is capable of being understood, at least in part. The physical universe exhibits prop-

2. We regard these questions as "difficult" for a variety of reasons: (1) they are often very complex, requiring the consideration of many diverse phenomena; (2) the events and processes under consideration cannot ordinarily be observed directly or repeated; (3) formulating answers is more dependent on plausibility arguments, or "postdiction," thereby reducing the degree of certainty that is attainable.

erties that are stable and measurable. And when we observe the behavior of physical systems, we discover patterns of physical behavior that appear to be universally present. By accepting this universality, we are able to make sense out of individual phenomena because they fit into larger patterns. Furthermore, because certain physical properties and patterns of physical behavior appear to be stable, we are able to recognize numerous features of the world around us as the products or consequences of earlier events and processes. Thus, we have discovered that even the formative history of the physical universe is intelligible. We are now able to reconstruct remarkably coherent scenarios for many past physical processes and events—each of them consistent with the basic patterns—that have made discernible contributions to the present state of affairs.

But why do we add the qualifier *intrinsic* when we say that the domain of natural science is limited to the intrinsic intelligibility of the physical universe? In essence, because we realize that the sciences cannot deal exhaustively with the physical universe. Although, as we noted earlier, the entire physical universe may be the object of scientific investigation, such investigation does not have access to every quality and/or aspect of the physical world. Only those qualities that are *intrinsic* (i.e., wholly resident within the empirically accessible physical universe) are included within the domain of scientific inquiry. Questions concerning other qualities and/or aspects of the universe and questions concerning the relationship of the physical universe to the rest of reality are of interest to the scientist as a whole person, but such questions do not lie within the limited domain of natural science. Natural science seeks to know the character of the component parts of the physical universe and their functional relationships to one another, but it sets aside the matter of the relationship of the physical world to any beings or realms of reality that transcend the physical world.

Questions concerning transcendent relationships lie outside of the domain of natural science. Because of limitations imposed by its methodology, science is unable, for example, to say anything about the relationship of the world to a divine Creator. Questions concerning the relationship of the universe or its history to God must be directed elsewhere. The incompetence of natural science relative to such matters must be honored by both theists and nontheists. Both must resist the temptation to coerce science into warranting (in the sense of proving) their particular religious beliefs.

Because of the importance of this distinction between the domains of intrinsic intelligibility and transcendent relationship, let us cite several instances in which the natural sciences, because of their limited domain of competence, must maintain respectful silence.

1. While natural science can fruitfully investigate the *formation* of various systems and structures within the physical world, it is incapable of dealing with the ultimate *origin* of the world's existence.

Because of a great deal of misunderstanding concerning the word *origin,* this statement must be clarified. As we noted in chapter 4, the word *origin* is often used as a substitute for the word *formation.* When geologists speak of the origin of the Grand Canyon, for example, they are concerned with the succession of physical events and processes that make up the formative history of this magnificent geological structure. The uplift of continental land masses and the process of fluvial erosion are examples of phenomena relevant to canyon formation. Similarly, when astronomers speak about the "origin" and evolution of planets or stars or galaxies, their concern is with the physical processes by which the present form of these celestial objects developed from earlier forms or structures. Even when cosmologists speak about the "origin" of the entire expanding universe, they speak in terms of those formative processes by which the present state of affairs developed from earlier states. In the context of purely scientific inquiry, a discussion of "origins" must necessarily be restricted to a consideration of the *formation* of physical structures within the universe, the existence of which is taken for granted.

The question of *ultimate origin,* however, goes far beyond the matter of formative history. When we ask about the ultimate origin of the universe we are asking about the source of the very existence of the universe, about what agent causes *something* to exist in place of *nothing.* Furthermore, it is important to note that we are asking not merely about the *beginning* of existence but about the existence of the universe at all times—past, present, and future—and even about the existence of time itself. The question concerning the ultimate source or cause or *origin* of the world's existence is just as much a question about right now as it is about any other moment in time.

This question of ultimate origin, however, lies well outside the domain of natural science. It is a question that demands a consideration of beings or agents or realms that transcend the physical universe. We are not saying that the question concerning the origin of the universe cannot be asked; we are only saying that any consideration of its answer takes us beyond the domain of natural science and into the domain of philosophy (metaphysics) or theology.

A diversity of answers has been offered. According to philosophical naturalism, for example, the universe is self-originating—that is, its existence is independent of any nonphysical creative agent. From the perspective of Christian theism, on the other hand, the origin of the world's existence is at all times dependent upon the active will of God, the Creator—just as dependent at this moment as at any other moment, even the first moment of time.

Questions of *origin*—of the ultimate source of existence itself—are profoundly important. Their answers, however, will never be derived from the results of natural science. They are religious questions that must be directed to whatever serves as the source of one's answers to religious questions. For Christians, that source is God's revelation through Scripture, the subject of careful consideration in chapter 7. Natural science, because of limitations in both the object and the domain of its investigation, has no choice but to remain silent.

2. While natural science can fruitfully investigate the *behavior* of the physical universe, it is incapable of settling the fundamental question concerning its *governance*.

As was the case for *formation* and *origin*, so also the distinction between *behavior* and *governance* is an important one—particularly in discussions concerning the relationship between science and religion. And, in a manner similar to our first distinction, we shall find one concept—*behavior*—to lie within the scientific domain, and the other—*governance*—to lie outside of its boundary.

When natural science investigates the behavior of a physical system, it is concerned principally with the empirically accessible physical processes that take place within that system or with physical interactions between that system and its environment. Geology, for example, is concerned with the behavior of the Earth's crust in response to processes occurring within the Earth itself and those that occur in response to Earth's interactions with the Sun and the Moon. Chemistry is concerned with the structure of atoms and with interactions among various atoms and molecules in a diversity of environmental conditions. Physics seeks to understand the behavior of physical systems and their interactions in terms of fundamental forces related to the physical properties of matter. And biology endeavors to understand the physical behavior of living systems in terms of the structure and behavior of the cell and its constituent parts and in terms of the interaction of an organism with its environment. In each case, natural science is concerned to describe the observable behavior of physical systems and to discover the general patterns of behavior into which any specific phenomenon can be placed.

The search for a comprehensive set of coherently interrelated general patterns lies at the heart of the scientific enterprise. Our descriptions of these universal patterns of physical behavior are known by various generic titles, such as "scientific theories," "theoretical models," and "laws of nature." Specific examples include the special theory of relativity, the kinetic-molecular model for gases, and the law of energy conservation. (Although some philosophers of science have chosen to do so, we make no hard and fast distinctions among the terms *theory, model,* and *law;* these terms are nearly interchangeable, and the association of any one of them with a particular concept is, we believe, more a matter of historical accident than of rigorous classification.)

For an illustration of the behavior/governance distinction, consider the law of energy conservation. According to this "law," all physical systems behave in such a way that the total amount of energy possessed by the system and its environment remains constant. Energy, we say, is always conserved; it can be neither created (in the sense of coming into being from nothing within the existing physical universe) nor destroyed—only changed in form or transferred from one system to another. The law of energy conservation is a remarkably useful statement describing a very important aspect of the behavior of physical systems. Natural science, by empirically investigating the behavior of a wide variety of physical systems, has discovered a certain regular pattern in physical behavior and has formulated the law of energy conservation to describe that behavior pattern.

But why does the physical world behave in accordance with that pattern or any pattern? What is the identity of the power or agent that *governs* physical behavior in a manner described by the energy conservation law? People sometimes speak as if the law itself governs that behavior; introductory textbooks are notorious for their talk about the "laws of nature that govern the behavior of physical systems." Such talk, however, is quite empty. The "laws of nature" are only our descriptions of the patterns of material behavior, and descriptions have no power to govern. The question of governance cannot be answered by describing patterns of behavior. Behavior patterns give evidence of a governing power at work, but such patterns are not themselves the source of governance. Behavior patterns are not the *cause* of governance; they are only the *result*.

Like the question of origin, the question of governance is fundamentally theological. We can illustrate this by noting the difference between the answers provided by two very different religious perspectives that are prominent in Western culture: philosophical naturalism (or materialism) and Christian theism. According to naturalism, there exist no transcendent beings; the physical world is all there is. The governance of material behavior must be performed by matter itself. Matter, according to naturalism, is self-governing—autonomous. Christian theism, on the other hand, identifies God as the Governor of physical behavior. What we customarily call the "laws of nature" are really our attempts to describe the patterns of divine governance. These are not laws *of* Nature for its self-governance but rather our working descriptions of what we perceive to be the intelligible manifestations of the effective will of God *for* the behavior of the created world. Physical behavior, according to theism, is not autonomous (self-governed) but theonomous (God-governed).

In light of this behavior/governance distinction, it should be evident that the proponents of naturalism and of theism need have no disagreements concerning the proper description for the patterns of physical behavior. Provided that they do their scientific work in conformity with the accepted standards for competence and integrity and that they employ the appropriate time-tested methods, people with vastly differing religious commitments can and do work together toward the common goal of faithfully describing the behavior of the physical universe—the object of scientific investigation.

However, while theists and philosophical naturalists need not disagree on matters of physical *behavior*, they are in profound disagreement on the matter of *governance*. But the choice between an autonomous or a theonomous perspective on the governance of physical behavior cannot be settled on the basis of scientific investigation. The proponents of these two different religious perspectives need not work toward the development of different and competing scientific descriptions of behavior, even though they seek to understand the governance of that behavior within the frameworks of very different religious perspectives. From the one perspective, matter is both self-existent and self-governing; from the other, God is the ultimate reality and the physical world is dependent on God for both its existence (origin) and its governance.

Locating the boundary of the domain of natural science is of crucial concern to those of us who wish to establish and maintain an amicable working relationship between science and theology. The approach taken in this book is based upon the recognition that while the *object* of investigation by natural science is the entire physical universe and its constituent parts—every physical thing that is empirically accessible—the *domain* of scientific inquiry is restricted to the intrinsic intelligibility of this universe. Working within this domain, the natural sciences are capable of investigating the remarkable degree of intelligibility that is resident within the physical universe itself—in its physical properties, in its patterns of physical behavior, and in its formative history.

On the other hand, questions concerning the relationship of the physical universe to any transcendent realm lie outside the scientific domain, within the domain of theological or philosophical (metaphysical) inquiry. Consequently, while natural science can deal fruitfully with the formative history of the universe, questions concerning the source of its existence must be directed elsewhere. Similarly, while questions concerning physical behavior are appropriate for scientific inquiry, the question concerning the identity of the governing agent must be recognized as theological. Questions of origin and governance must be directed toward whatever serves as the source of answers to one's theological questions. On such matters the natural sciences have nothing to contribute.

The Values Operative in Scientific Praxis

The "epistemic" goal of the natural sciences is to gain knowledge of the intrinsic intelligibility discernible in the physical properties, behavior, and formative history of the physical world in which we live. The principal forms of this knowledge are the results of empirical investigation and the products of scientific theorizing about the composition, structure, behavior, and history of the physical systems that we observe.

In the remainder of this chapter we focus our attention on the system of *values* that functions in the process of scientific theorizing. Other aspects of theoretical science could have been selected. Students of the natural sciences, for example, must spend a large amount of time becoming familiar with the *content* of numerous theories. And the developers of modern technology expend the majority of their effort in the creative and effective *application* of relevant theories to the accomplishment of certain practical goals. But professional natural scientists—people who are engaged in basic scientific research—are most deeply concerned with the formulation and *evaluation* of theories about physical phenomena.

Judging the merits of a particular scientific theory or choosing one theory from a set of competing theories concerning the same phenomena is a common activity within the scientific community.[3] But on what basis

3. By *scientific theory* we ordinarily mean a set of concepts concerning the physical properties, physical behavior, and sometimes the formative history of some physical system or class of systems and the way in which these aspects of physical systems are interrelated (e.g., by proximate cause-effect relationships). We implicitly recognize that such a set of concepts, while it may systematically organize and enhance our understanding of the physical nature of something, represents an abstraction from the full reality of that entity and does not exhaustively account for all of its qualities or functions.

are such judgments and choices made? A half century ago it was common to suppose that there existed some set of self-evident and rigidly applicable rules by which the truth or falsehood of a particular theory could be established once and for all. During the past few decades, however, historians, sociologists, and philosophers of science have been able to develop a more realistic assessment of the way in which the scientific community functions. It is now generally agreed that scientific theorizing is very little like the positivist picture of the mechanical application of rigid logical *rules;* rather, it is a *value*-guided activity of human judgment applied to the products of creative insight. Scientific theories cannot be constructed by a robotic assembly line, nor can theory evaluation be adequately performed by a room full of the most powerful of today's electronic computers.

If we wish to gain some insight into the process of scientific theory evaluation, we must see it as a wholly human enterprise. It is an activity performed by a community of individuals—people with finite knowledge, skills, and insights—who must continually make judgments concerning the adequacy of scientific theories to account for the results of empirical investigation. Judgments must, of necessity, be based on accepted standards or values.

There exists no authoritative document that spells out the criteria for acceptable scientific investigation, not even for the more restricted activity of theory evaluation. In the absence of such a canonical source, the best we can do is to appeal to scientific practice itself in order to determine what general principles or value systems appear to be operative. What follows, therefore, is our summary of insights into the historical practice of natural science provided by several observers of the scientific enterprise.[4] Our list is by no means exhaustive. We have limited

---

As we use the term, *theory* is not to be contrasted with either "fact" or "hypothesis." These latter terms refer to the degree of certainty with which a particular theory may be judged an adequate means of describing some physical system or phenomenon. As here defined, scientific theories are always subject to evalution, revision, or replacement as the scientific community functions to develop an increasingly adequate array of coherently interrelated theories.

4. See, for example, the following: Thomas S. Kuhn, *The Structure of Scientific Revolutions,* 2nd ed. (Chicago: University of Chicago Press, 1970), and *The Essential Tension* (Chicago: University of Chicago Press, 1977), especially chap. 13, "Objectivity, Value Judgment, and Theory Choice," pp. 320-39; Larry Laudan, *Science and Values* (Berkeley and Los Angeles: University of California Press, 1984); Ernan McMullin, "Values in Science," *PSA 1982: Proceedings of the 1982 Biennial Meeting of the Philosophy of Science Association,* vol. 2, ed. Peter D. Asquith and Thomas Nickles (East Lansing, MI: Philosophy of Science Association, 1982), pp. 1-25; and Jerome R. Ravetz, *Scientific Knowledge and Its Social Problems* (New York: Oxford University Press, 1971).

ourselves to brief discussions of four categories of values, four categories of functioning criteria for judging the quality of scientific research or the adequacy of a scientific theory.

Matters of Competence

Scientific research and theorizing comprise a rich diversity of actions performed by scientists. Examples abound: a long-range research program is planned, motivated by factors as varied as the colors in a rainbow; a specific question is posed for investigation; the investigators thoroughly familiarize themselves with the relevant professional journal literature; an empirical strategy of investigation is formulated; apparatus is selected or designed and assembled; the physical system or sample of material to be investigated is prepared; arrangements are made to control the environmental conditions; the measurement system to be used is calibrated against an accepted standard; the degree of uncertainty in measured values is assessed; the observations or measurements of interest are performed and the relevant data are recorded (measurements may be repeated); the data are analyzed; the results are organized for presentation in the form of verbal descriptions, tables of numerical values, graphs, algebraic relationships, or the like; the results may be compared with predictions or expectations; inferences concerning relevant theories are drawn, or a new theory may be proposed to account for the empirical results; suggestions for further related research are offered, perhaps leading to a similar cycle of empirical and theoretical activity.

Each of these activities constituting scientific research requires a high degree of familiarity with appropriate, tested procedures and competence in performing them. To be a good natural scientist, one *must* acquire the craft knowledge and the skills that are needed to perform these empirical, analytical, and theoretical operations fruitfully. Craft competence is highly valued within the professional scientific community. The tradition of high expectations for competent performance is passed on from one generation of scientists to the next. Incompetent work is universally rejected.

But surely this demand for competence comes as no surprise. After all, the epistemic goal of natural science is authentic knowledge concerning the character of the physical universe. The only paths leading toward that goal are those paved with the results of competent exercise of the empirical and theoretical crafts; good intentions are not enough. Incompetent performance in scientific investigation leads only to misperception, delusion, and false conclusions—no friends of knowledge. But, of course, incompetent performance is unwelcome in any field of endeavor.

Matters of Integrity

Scientific investigation is performed by a community of people who must depend upon the professional integrity of the other members of that community. While it may often be the case that research performed by one person or group will, for a variety of substantial reasons, be challenged and repeated by another, the more common case is that research reports are trusted to represent the results of competent work honestly reported. The *significance* of the report is open to question, and the *convincing power* of its argumentation is subject to critical evaluation, but its *integrity* is expected to meet the unwritten but nonetheless functional code of professional ethics. Without a functioning set of ethical principles of integrity, the professional scientific community could not perform in the manner it presently does.

The fundamental principles are honesty, fairness, and candor. Unquestionably, the willful propagation of reports that are known to be false or unreliable is totally unacceptable. And when observations or measurements are reported, one expects not only that the report provides an honest account of the results but also that all reasonable precautions have been taken to ensure that these empirical results are reliable within limits that have been realistically assessed and candidly stated. And if in the course of a computation one needs to use data obtained by other researchers, it is expected that the relevant scientific literature has been thoroughly searched in order to obtain the most reliable data available (often the data most recently reported). Intentional use of unreliable or discredited reports in support of one's case is universally deplored. Professional scientists are expected to exercise whatever level of diligence and self-discipline is required to minimize the propagation of false or misleading reports.[5]

The process of extrapolation deserves special attention in the context of this discussion of professional integrity. It is ordinarily the

5. For several examples of the mischief that is perpetrated when the expectations of professional integrity are disregarded, see Howard J. Van Till, Davis A. Young, and Clarence Menninga, *Science Held Hostage: What's Wrong with Creation-Science AND Evolutionism* (Downers Grove, IL: InterVarsity Press, 1988), pp. 45-124.

case that the data base for the description of some behavior pattern is confined to a restricted range of circumstances. Any extrapolation of behavior patterns beyond those limits must be performed with appropriate restraint, and the conditions for the credibility of that extrapolation must be candidly stated. If such restraint is not exercised with integrity, the results are likely to be meaningless at best, and they would be grossly misleading for persons unprepared to assess the credibility of a given extrapolation. Mark Twain provides us with a colorful example of the problematic conclusions that can be drawn from unrestrained extrapolation:

> In the space of one hundred and seventy-six years the Lower Mississippi has shortened itself two hundred and forty-two miles. This is an average of a trifle over one mile and a third per year. Therefore, any calm person, who is not blind or idiotic, can see that in the Old Oolithic Silurian Period, just a million years ago next November, the Lower Mississippi River was upward of one million three hundred thousand miles long, and stuck out over the Gulf of Mexico like a fishing-rod. . . . There is something fascinating about science. One gets such wholesale returns of conjecture out of such a trifling investment of fact.[6]

Professional science concurs wholeheartedly with the thrust of Twain's conclusion: unrestrained extrapolation produces misleading nonsense.

Several matters of integrity are involved also in the manner in which one argues in favor of a particular theoretical model. We expect, for instance, that all relevant data will be given fair and adequate treatment—not only the supportive data but also the data that stand in tension with the model. Anomalous or contradictory evidence may not be neglected or concealed. And if there are other phenomena or theories relevant to the one under evaluation, we expect the writer of the research report to bring such matters to the attention of the reader and to demonstrate how they strengthen or weaken the case. This is particularly important when arguing in favor of a theory that makes strong claims for its superiority over a conventionally accepted one. Thus, in addition to the specific data and phenomena that are being investigated, the context of the investigation provided by other relevant data, phenomena, and theories must be honestly engaged. Theorizing that by conscious choice or by gross neglect fails to pay due attention to this context also fails to meet the standards of professional integrity that are necessary for the proper functioning of the scientific community.

6. Twain, *Life on the Mississippi* (New York: Harper & Brothers, 1917), p. 156.

Matters of Sound Judgment

Not infrequently it will happen that two or more competing theories claim to account for relevant data and provide the best explanation for a given phenomenon. In such cases, scientists have to judge the relative merits of the competing theories and make choices on the basis of certain criteria. Assuming that the criteria for competent research and honest analysis have already been satisfied, what additional criteria does the scientific community employ in its evaluation of competing theories? Following the suggestion of Ernan McMullin, we refer to these as *epistemic values,* the criteria used to evaluate the scientific merit of a theory—that is, the likelihood of its being authentic knowledge concerning the actual state of affairs.

As in the case of professional ethical values, there exists no standard list of epistemic values. No professional scientific organization has undertaken, nor is any likely to undertake, the formulation of a program for scoring or grading a theoretical model. That's not the way value judgments are made. However, for the purpose of illustrating how the scientific community does go about the business of theory appraisal, let's list examples of relevant epistemic values. McMullin cites six—predictive accuracy, internal coherence, external consistency, unifying power, fertility, and simplicity.[7] Our list is a modified version of his. Several examples of how these criteria function in practice will be provided in the concluding section of this chapter.

1. *Cognitive relevance.* The supply of diverse theories concerning some aspects of the composition, structure, behavior, or history of the physical realm appears to be inexhaustible. But many of these theories (especially those proposed by people who are not specialists in the area under consideration) fail to be of any value to natural science because they do not have cognitive relevance—that is, they fail seriously to engage the relevant empirical evidence. For example, college science teachers often receive personal letters or copies of privately published papers that claim to present grand theories that "solve the great riddles of the universe" and soon, purportedly, will replace the whole array of conventional scientific theories. These proposals, usually written by individuals with little scientific training, have as one of their characteris-

7. McMullin, "Values in Science." For other discussions of scientific theory evaluation, see W. H. Newton-Smith, *The Rationality of Science* (Boston: Routledge & Kegan Paul, 1981), pp. 226-32; Kuhn, *The Essential Tension,* pp. 320-39; and Del Ratzsch, *Philosophy of Science: The Natural Sciences in Christian Perspective* (Downers Grove, IL: InterVarsity Press, 1986), pp. 75-96.

tic features the absence of meaningful contact with specific empirical data relevant to the phenomena under discussion. References to actual physical properties or behavior are typically very general and vague at best. Such proposals, because of their lack of engagement with actual data, are of no use in the search for knowledge. They may themselves provide the data for an interesting psychological study, but they contribute nothing to our understanding of the physical world.

In addition to these occasional contributions to the "psychoceramic file" (the receptacle for crackpot literature), there are numerous other theories about the physical world, especially about its formative history, that have been constructed with little or no regard for the actual physical data that are readily available in the scientific literature. The impetus for these theories may be the desire to reinforce some philosophical perspective or to provide support for some religious tenet. The elements that constitute the theory or historical scenario may be drawn from sources or traditions considered to be relevant and authoritative by a particular ideological community, or from imaginative speculation, or from both. However, even when such theories are proposed by intelligent, well-educated, and sincere people, they provide little or no assistance in achieving the epistemic goal of natural science. A theory that fails to engage the relevant empirical data will not be able to provide a coherent picture of the world as it really is. Natural science seeks knowledge concerning the intrinsic intelligibility of the world in which we actually live. Its theories must fully engage the results of empirical investigation or be judged to have little epistemic value.

2. *Predictive accuracy.* As a general rule, an authentically scientific theory provides a means of predicting the values of measurable properties or the character of certain behavioral characteristics that a given physical system should exhibit. The obvious question to ask is how well the predicted quantities or behavior patterns compare with those observed. In assessing the merit of one or more theoretical models, the criterion of predictive accuracy is surely relevant. A model that displays greater predictive accuracy than some competing model would ordinarily be favored. There is, however, no absolute guarantee that the model favored on the basis of this single criterion will in the long run prove to be the better one. Ptolemy's geocentric model for the planetary system may have had greater predictive accuracy than the heliocentric model suggested earlier by Aristarchus, but we now know that the heliocentric model, as improved by Copernicus, Kepler, and Newton, has clearly demonstrated its superiority in a number of different ways, including predictive accuracy.

A slight variation on this criterion is applicable to the investigation of formative history. In this case our concern is not to predict what will happen in the future but rather to construct a plausible scenario for what did happen in the past, thereby giving a specific form to what we are now able to observe. The relevant question to ask of a formative history model is how well its historical reconstruction accounts for the present state of affairs. Would the proposed formative history lead us to expect all of the features that have already been observed? Would the model also predict the presence of features not yet investigated? And when investigators look for these predicted features, are they found to be there? If so, one's confidence in the credibility of that formative history scenario is legitimately increased.

3. *Coherence.* To those of us who have grown up in the environment of twentieth-century Western culture, it is perhaps self-evident that an adequate scientific theory should be internally coherent, that it should contain no elements that are inconsistent with other elements. We assume that the behavior of physical systems is internally consistent, and consequently we expect that our theoretical models for their behavior will be devoid of any internal contradictions.

But the criterion of coherence has an even broader scope. We expect not only that the behavior of a particular system or category of systems will be internally coherent but also that the physical behavior of the entire empirically accessible universe will be coherent in the sense of entailing no inconsistencies or contradictions. Patterns of physical behavior are presumed to be universally applicable—the same patterns in all places and at all times and for all relevant systems. The law of energy conservation, for instance, applies not only here on Earth but within the Andromeda Galaxy as well, not only today but three billion years ago in the quasar 3C 273 as well, not only for falling apples but for nuclear reactions as well. In the context of such expectations, then, we would ordinarily judge that a scientific theory should display not only an internal coherence but also a coherence with respect to the entire spectrum of physical phenomena and their associated theoretical models. The adequacy of any theory will in part be judged on the basis of its coherence relative to other theories already shown to have epistemic merit.

4. *Explanatory scope.* The scientific community does not rest comfortably with unexplained coincidence. To cite a classic example, the Ptolemaic geocentric model for the solar system left two specific features of the apparent motion of Mercury and Venus without explanation. When we observe these two planets from Earth, they always appear to

be relatively close to the Sun. Expressed in terms of angle, Mercury remains within twenty-five degrees of the Sun's apparent position in the sky, and Venus within fifty degrees. When these planets appear westward of the Sun's position they rise shortly before sunrise and we see them above the eastern horizon as "morning planets." At other times, when the planets appear eastward from the Sun, we see them as "evening planets" in the western sky for a brief period after sunset. Employing the language of the Ptolemaic theory, we would conclude from these observations that the epicycle centers for these planets was constrained to lie on the Earth-Sun line. Why should their motion satisfy this constraint? Within the Ptolemaic model there was no explanation; the constraint stood as an unexplained coincidence. A similar appraisal applied to the observation that the fastest retrograde motion (westward relative to the background of stars) exhibited by Mars, Jupiter, or Saturn always occurs when the planet is in opposition to the Sun—another unexplained coincidence within the framework of Ptolemy's Earth-centered geometry. In the Sun-centered models of Copernicus and Kepler, however, both of these phenomena came to be recognized as natural consequences of the heliocentric geometry of the models. The explanatory scope, or inclusiveness, of the Sun-centered models was greater than that of the Ptolemaic geocentric model. Unexplained coincidences were transformed into natural consequences.

Modern examples could also be cited. For instance, the fact that the chemical composition of the physical universe is dominated by hydrogen and helium in a three-to-one ratio by mass is no longer considered to be a curious accident but rather the natural consequence of early cosmic history. The ability of the "standard model" in contemporary cosmology to explain this feature is considered to be a strong point in its favor. Theories with greater explanatory scope are generally judged to have greater epistemic merit.

5. *Unifying power.* Beyond its attempt to develop theories with sufficient explanatory scope to eliminate unexplained coincidences, the scientific community also seeks to unify what may once have been viewed as unrelated phenomena, each explained by means of independent theories, into a single, more comprehensive theoretical framework. The contemporary effort by elementary particle theorists to develop a single theory that will encompass the four (or will it be five?) fundamental forces is a good example of this.[8] In the nineteenth century, Maxwell

8. By "elementary particles" we mean the basic building blocks of which physical systems are made. An atom, for instance, is composed of an atomic nucleus surrounded

successfully unified electrical and magnetic phenomena with his theory of electromagnetism. Today's theoretical physicists seek to develop a theory that will allow us to view the electromagnetic force, the weak nuclear force, the strong nuclear force, and the gravitational force as but different manifestations of a single "superforce."[9] Remarkable progress toward this unification of forces appears already to have been achieved.

But even on scales far less grand than that envisioned for a "superforce" theory, those scientific theories that are able to unify diverse phenomena into one comprehensive theoretical framework will be favored over a collection of independent theories that treat each phenomenon in isolation from the others. The greater the unifying power of a theory, the better able it is, we judge, to demonstrate the intrinsic intelligibility of the physical world.

6. *Fertility.* The criterion, or epistemic value, of fertility is a bit more elusive than those already discussed. The five criteria cited above are concerned primarily with the question of how well a given theory accounts for what we already know. The criterion of fertility, on the other hand, is concerned with how well a theory functions to stimulate investigation in new areas, to suggest new ways of organizing our knowledge, to reveal relationships previously obscured, and the like. In each case the quality under scrutiny is the potency of a theory for stimulating the imagination and initiating the propagation of a continuing line of helpful insights.

MORE EPISTEMIC VALUES (simplicity and "beauty," for example) could be cited, though perhaps with diminished consensus among the philosophers and practitioners of science concerning their importance within the scientific enterprise. Among those cited above there is unavoidably some overlap. Furthermore, although we chose to arrange our list in the way we did, we make no claims that this represents the order of their functional importance. With McMullin, we intend only to call attention to the idea that theory assessment and theory choice in natural science

---

by electrons in motion. The nucleus is made up of protons and neutrons, and the protons and neutrons are each composed of "quarks." Electrons and quarks are, we believe, examples of *elementary particles.* Interactions among particles are described in terms of *fundamental forces,* of which four are known: gravity, the electromagnetic force, the "weak" nuclear force, and the "strong" nuclear force. Some theorists have proposed the existence of additional fundamental forces, and empirical investigation is under way to test these hypotheses.

9. See Paul Davies, *Superforce* (New York: Simon & Schuster, 1984).

is not principally a matter of mechanically scoring a theory according to some fixed set of self-evident rules or merely a matter of expressing one's personal opinion; it is more like the process of making value judgments, something everyone does every day. These judgments are based on the system of *epistemic values* resident within the scientific community—values inherited from previous generations, modeled by senior members of the community, strengthened or modified by experience, applied in varying ways by the several members of the community—and their normal employment permits a healthy level of disagreement within the context provided by a broad foundation of consensus. The epistemic values now functioning are the time-tested products of reflection on organized experience.

For the sake of emphasis, it is worth repeating an earlier comment that there is "no authoritative document that spells out the criteria for acceptable scientific investigation." The long research process that culminates in the publication of an acceptable piece of scientific work involves innumerable judgments involving *craft competence* (e.g., When is a data set good enough to use? When is a piece of apparatus working well enough? What is the best technique for preparing a specimen? etc.), *professional integrity* (Has an adequate literature search been performed? Have conflicting results been acknowledged and adequately discussed? Have possible sources of error been recognized and realistically assessed? etc.), and the several *epistemic criteria* discussed above. Together, these three categories of values play a major role in shaping the broad outlines of what counts as evidence, information, and knowledge within the natural sciences. The hallmark of a professional scientist is the ability to consistently make appropriate judgments that reflect high levels of craft competence, professional integrity, and epistemic discernment.

But in many (perhaps most) instances, explicitly stated criteria are not available for use in deciding whether or not a data set is "good enough to use" or if an "adequate literature search has been performed" or whether a proposed theory is coherent "with respect to the entire spectrum of physical phenomena." Nevertheless, scientists are able to make such decisions using *tacit* criteria, and their peers are able to recognize when a research project embodies the competent exercise of scientific judgment. Tacit criteria are learned gradually by neophyte researchers through imitation and through the continuous round of discussion and criticism that is of central importance in strong graduate programs. Within an established field of research, a broad consensus can be sustained as its members engage in a continuous dialogue employing numerous avenues of formal and informal communication. Within such a community the role of senior members is particularly important. Their published research exemplifies the current state of the art, and their actions as journal editors, reviewers, and members of panels reviewing grant applications, as well as their work as teachers and advisors, provide many opportunities to enforce their convictions concerning competent performance.[10]

Extrascientific Matters

We have called attention to several constraints on the scientific enterprise that have evolved within the community of practicing scientists. The *object* of scientific investigation is not all of reality but only the physical world. The *domain* of natural science does not encompass all categories of questions about the physical world but only those questions concerning its intrinsic intelligibility. Operating within the boundaries of that domain, members of the scientific community are further constrained by their professional colleagues to meet the community's standards for craft competence, to carry out and report their research with ethical integrity, and to evaluate scientific theories on the basis of a communally developed epistemic value system. But what, we now ask, do scientists do with their concepts and beliefs concerning the rest of reality? Must their scientific work be completely isolated from their concerns in the arenas of religion, philosophy, politics, economics, social institutions, personal ambitions, the arts, and the like? Will not these cultural, ideological, personal, religious, and other extrascientific concerns—concerns for matters outside of the scientific domain—have a discernible influence on their scientific work? If so, what is the character and extent of this influence?

Identifying the roles played by extrascientific concerns in the scientific enterprise is a complex and difficult task. It is an area of study

10. See the discussion of tacit knowledge in Ravetz, *Scientific Knowledge and Its Social Problems*, pp. 101-3. Ravetz notes that "The scientist's craft also includes the formulation of problems, the adoption of correct strategies for the different stages of the evolution of a problem, and the interpretation of general criteria of adequacy and value in particular situations. These . . . tasks, which distinguish original scientific work from the routine production of bits of information, have no standardized, elementary versions to which simple, explicit precepts can apply. Hence most of the body of methods governing this work is completely tacit, learned entirely by imitation and experience, perhaps without any awareness that something is being learned rather than 'common sense' being applied" (p. 103).

that deserves continuing attention. For our present purposes we shall restrict our remarks to a few brief comments on the function of religious commitments in scientific investigation.[11] The term *religious commitment* must not be interpreted too narrowly, however. We do not wish, for instance, to restrict it to the Christian religion, nor even to theism in general. Rather, we intend the term to represent the full spectrum of beliefs concerning the ultimate nature of reality, the existence or nonexistence of a transcendent deity, the significance of human life, and the relationship of the physical world to any transcendent beings or realms of reality. That said, we can see that evolutionary naturalism is as much a religious orientation or commitment as Christian theism.

1. Religious commitments frequently serve as a stimulus for a scientist to select and carry out a particular program of research. Certain topics for investigation may be given priority because of their relevance to an investigator's worldview—his or her "vision of reality." We see no reason to criticize or discourage this kind of influence. On the contrary, it is a wholly appropriate way to act with integrity in the context of one's religious perspective.

2. Religious commitments ought never to lead a scientist to permit or encourage any reduction in the demands for craft competence or professional integrity. In order for the epistemic goal of natural science—the gaining of authentic knowledge—to be achieved, each member of the community must honor the requirements for competence and integrity and must participate in the process of mutual discipline that functions to maintain those standards.

3. Religious commitments cannot be used as a warrant to ignore or to consciously violate the boundaries of the scientific domain. While the domain of one's personal concerns will inevitably extend beyond the boundaries of natural science, no scientist has the right to claim that natural science itself has the competence or authority to settle issues outside the domain of intrinsic intelligibility that we described earlier. For example, the oft-heard claims that natural science either "scientifically" confirms or discredits a theistic concept of divine governance or "scientifically" validates some particular concept of the status of the physical universe in relationship to deity is careless talk that exposes a failure to recognize the boundaries of the scientific domain. Such a mischievous violation of domain boundaries is likely to be damaging to the credibility

11. For a discussion of the role of a Christian worldview in conducting scientific investigation, see Howard J. Van Till, "Scientific World Pictures within the Bounds of a Christian World View," *Pro Rege*, March-June 1989, pp. 11-18.

of authentic scientific results and, in the long run, will do a particular religious perspective no favor. Linking a specific scientific theory with some religious belief system in such a way that one entails the other, for example, has a serious strategic disadvantage in that any discrediting of that scientific theory automatically tends to call into question the entire belief system attached to it.[12] Alliances of religious belief and scientific theory should be formed only with great care and restraint and in a manner that promotes an openness to continuing discovery.

4. Religious commitments, whether theistic or nontheistic, should not be permitted to interfere with the normal functioning of the epistemic value system developed and employed within the scientific community. Great mischief is done when extrascientific dogma is allow to take precedence over epistemic values such as cognitive relevance, predictive accuracy, coherence, explanatory scope, unifying power, and fertility. And progress toward the goal of authentic knowledge is likely to be impeded when religious commitment is permitted to so skew the theory evaluation process that one epistemic value takes inordinate precedence over all others.

The troublesome tendency with which we are dealing here is the temptation to employ natural science for the purpose of supporting preconceptions about the physical universe drawn from one's philosophical commitments or system of religious beliefs. Such an approach stands the scientific enterprise on its head and must be resolutely avoided. The goal of natural science is to gain knowledge, not to reinforce preconceptions. The purpose of empirical research is to discover what the physical world as an intelligible system is really like, not to verify its conformity to our prejudices. And the aim of scientific theoriz-

12. A closely related problem arises when one demands, for religious reasons, a belief about the physical universe that has already been soundly discredited by scientific investigation. Augustine speaks to us about the tragic consequences of such an approach:

> Usually, even a non-Christian knows something about the earth, the heavens, and other elements of this world, about the motion and orbit of the stars and even their size . . . , and this knowledge he holds to as being certain from reason and experience. Now, it is a disgraceful and dangerous thing for an infidel to hear a Christian, presumably giving the meaning of Holy Scripture, talking nonsense on these topics; and we should take all means to prevent such an embarrassing situation, in which people show up vast ignorance in a Christian and laugh it to scorn. . . . If they find a Christian mistaken in a field which they themselves know well and hear him maintaining his foolish opinions about our books, how are they going to believe those books in matters concerning the resurrection of the dead, the hope of eternal life, and the kingdom of heaven . . . ? (*The Literal Meaning of Genesis*, 2 vols., Ancient Christian Writers, nos. 41-42, trans. John Hammond Taylor [New York: Newman Press, 1982], 1:42-43)

ing is to describe the actual character of the universe, not to force its compliance with our preconceived requirements.

Science held hostage to any ideology or belief system, whether naturalistic or theistic, can no longer function effectively to gain knowledge of the physical universe. When the epistemic goal of gaining knowledge is replaced by the dogmatic goal of providing warrant for one's personal belief system or for some sectarian creed, the superficial activity that remains may no longer be called natural science. One may call it "worldview warranting," "creed confirmation," or "apologetic science," or one may put it into the category of "folk science,"[13] but it no longer deserves the label of "natural science" because it is no longer capable of giving birth to authentic scientific knowledge. Science held hostage by extrascientific dogma is science made barren.

Are we then left with the implication that religious belief is held hostage by the results of an autonomous natural science? Emphatically not![14] Because, as we have seen, science and religion have different domains of competence and concern, each needs to learn from the other concerning what lies outside of its own domain. And those of us who wish to build a comprehensive worldview must learn from both so that we may come to know not only the intrinsic intelligibility of the physical universe but also its place within the whole of reality. The scientific community, made up of people representing a diversity of religious perspectives, seeks to gain knowledge of the physical universe in a manner appropriate to the object of its specialized investigation. When individual scientists go beyond the scope of their special sciences and speak on matters concerning the ultimate meaning or significance of the composition, structure, or formative history of the universe, they may

13. We are using the term "folk science" in a manner similar to that of Ravetz in *Scientific Knowledge and Its Social Problems*, especially pp. 386-97. Ravetz defines folk science as that "part of a general world-view, or ideology, which is given special articulation so that it may provide comfort and reassurance in the face of the crucial uncertainties of the world of experience" (p. 386). For an explanation of the relationships linking science, folk science, and worldview, see our chap. 6, pp. 186-91.

14. Numerous books are available that document the fruitful interaction of natural science and Christian belief. As a start, we recommend the following: R. Hooykaas, *Religion and the Rise of Modern Science* (Grand Rapids: Eerdmans, 1972); N. Max Wildiers, *The Theologian and His Universe* (New York: Seabury Press, 1982); David N. Livingstone, *Darwin's Forgotten Defenders* (Grand Rapids: Eerdmans, 1987); and *God and Nature: Historical Essays on the Encounter between Christianity and Science*, ed. David C. Lindberg and Ronald L. Numbers (Berkeley and Los Angeles: University of California Press, 1986). For more popular discussions, see Colin Russell, *Cross-Currents: Interactions between Science and Faith* (Grand Rapids: Eerdmans, 1985), and Charles E. Hummel, *The Galileo Connection: Resolving Conflicts between Science and the Bible* (Downers Grove, IL: InterVarsity Press, 1986).

disagree sharply because of their different religious commitments. However, when working within the restricted domain of natural science, scientists are able to function as a community united in the search for authentic knowledge concerning the intrinsic intelligibility of the physical world. The results of professional natural science belong to everyone; any attempt to declare them to be the exclusive property of one specific religious perspective must be summarily rejected.

STAR CLUSTERS: AN ILLUSTRATIVE EXAMPLE

Some Puzzling Observations

More than half of the observable stars are members of multiple-star systems. Many of these systems are simple binaries—two stars in mutual orbit about a common point. But some of these stellar collections are made up of a larger number of stars clustered in a fairly compact arrangement.

Two distinct types of clusters are observed: *galactic* clusters and *globular* clusters. Galactic clusters, such as the Pleiades, usually contain a few hundred stars and are located near the plane of the galaxy to which they belong—among the stars and nebulae that make up the spiral arm structure. Globular clusters, so named for their spherically symmetric structure, are composed of several hundreds of thousands of stars and are ordinarily found outside of the galactic plane in the "halo" region—a spherical volume centered on the galactic center and having a diameter comparable to the diameter of the galactic disk. M 13, the thirteenth entry on Messier's eighteenth-century list of fuzzy patches of light that were not comets, is a magnificent globular cluster that can be seen in the direction of the constellation Hercules, visible to northern hemisphere observers in the summer sky in early evening. Our home galaxy, the Milky Way, contains thousands of galactic clusters and at least 125 globular clusters.

Although galactic clusters and globular clusters differ in several of their properties (number and distribution of stars, location, etc.), our concern here is to focus on one particular feature that clusters of both varieties have in common. This is a feature displayed by the distribution of the stars of a cluster on an H-R diagram—not their spatial distribution, a matter of star locations, but their distribution on the diagram based on stellar luminosity and temperature values.

Recall from chapter 4 that a random sample of stars is observed

Above is galactic cluster NGC 2682; below is NGC 6205, a globular cluster in the constellation of Hercules. Our Milky Way Galaxy contains thousands of galactic clusters and at least 125 globular clusters. Palomar Observatory Photographs

to contain a mixture of main-sequence stars, red giants, and white dwarfs distributed on a luminosity-temperature diagram as illustrated on p. 87. However, when we measure the values for the surface temperature and luminosity of the stars constituting a representative cluster and place these results on an H-R diagram, a remarkable and intriguing peculiarity appears. As shown in Figure 1 on page 154, most of the stars in a representative cluster are main-sequence stars. But a careful inspection of the diagram reveals two very puzzling features that should arouse our curiosity: (1) a portion of the upper end of the main sequence is unpopulated, and (2) there are some stars distributed in an appendage to the main-sequence region, extending upward and to the right from the upper end of the populated portion of the main sequence. Because this peculiar distribution appears to be turning off from the normally occupied area, the point on the diagram at which this appendage leaves the main-sequence region is called the *turnoff point.* For all globular clusters the turnoff point is located near the Sun's position on the main sequence. For galactic clusters the turnoff point is usually higher up, in the vicinity of the more massive, shorter-lived stars.

Many additional details could be supplied with this description of cluster diagrams, but what we have selected is adequate for the immediate illustrative purposes of this discussion.[15]

Formulating a Solution

Clearly the H-R diagrams for star clusters exhibit an intriguing peculiarity that distinguishes them from the diagram for any random sample of stars. But such puzzling results of observation provide the principal motivation for creative scientific theorizing. Concerning star clusters, the results of careful and competent observations raise several specific questions that need to be addressed: Why is the upper end of the normal main-sequence region not populated with stars? Why instead are there stars distributed along a strip from the turnoff point toward the red giant region? And why is the location of the turnoff point the same for all globular clusters, but not for galactic clusters? In sum, *what leads star clusters to exhibit this particular set of peculiar features?*

Before looking at the way in which astronomical theorizing has

15. For textbook discussions of star clusters, see George O. Abell, *Exploration of the Universe,* 4th ed. (New York: Saunders College Publishing, 1982), pp. 492-502; and Frank H. Shu, *The Physical Universe* (Mill Valley, CA: University Science Books, 1982), pp. 159-78.

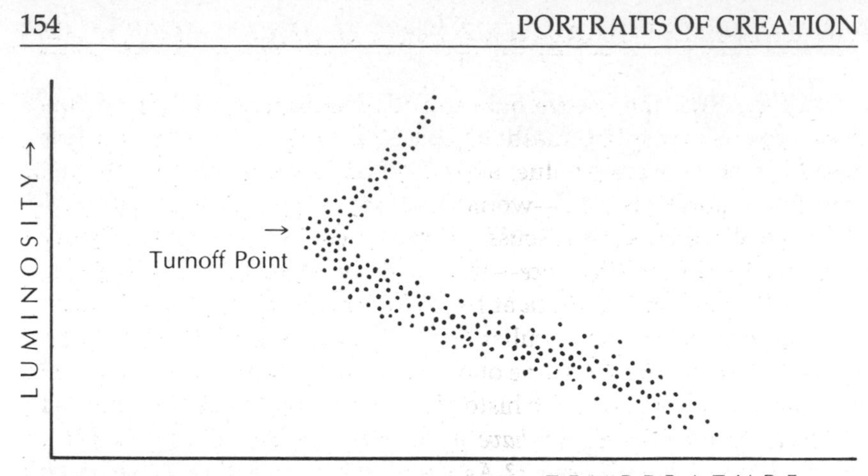

Figure 1. H-R Diagram of the Stars in a Representative Stellar Cluster. A comparison of this diagram with the H-R diagram for a random sample of stars (see p. 87) reveals a significant difference: on the cluster diagram, the portion of the main sequence that lies above and to the left of the turnoff point is unpopulated—the cluster contains no massive stars in their main-sequence phase.

proceeded to construct and evaluate a solution to this puzzle, we ought explicitly to disqualify one particular strategy. In answer to the request, "Tell me why the stars do shine," a song familiar to some of us answers, "Because God made the stars to shine." In a *theological* sense, this reply is, we believe, correct. It is *not,* however, an adequate *scientific* answer to a legitimate *scientific* question about stellar energy generation. Neither should it be viewed as an appropriate alternative or substitute for a scientific answer. To say that God makes stars shine is to speak theologically to questions regarding the governance of all physical phenomena. But to identify God as Governor still provides no help at all in understanding the specific physical processes involved in the generation of luminous energy by a heated stellar surface. To say "God makes the stars to shine" gives us no aid in scientifically apprehending the *intrinsic intelligibility* of the stars in his Creation. Similarly, to answer the question "Why do star clusters exhibit these particular peculiarities?" by saying "Because that's the way God made them" is scientifically inadequate. The nontheistic version of this answer would be "Well, that's just the way star clusters happen to be!" In either case we would be saying that the special features that are exhibited by the H-R diagrams for clusters are nothing more than amusing coincidences, and no further questions should be asked. In the context of what we consider to be the healthy attitude of contemporary scientific curiosity, such responses are nothing short of absurd—they serve only to stifle learning and to perpetuate ignorance. As replies to questions of a scientific nature, answers like these have no epistemic value, and responsible scholarship—whether Christian or non-Christian—would dismiss them in an instant.

Recall from earlier discussion that a meaningful scientific theory must have *cognitive relevance*—that is, it must relate as directly as possible to the empirical data at hand. In our example of star clusters, we have focused on the peculiar features exhibited by their H-R diagrams. But to enlarge the scope of our consideration, we would do well to inquire about the formative history of clusters. *Is it possible,* we should ask, *that the peculiarities we have highlighted are the consequences of a particular historical development?* As we soon shall see, this has been an especially fruitful line of inquiry.

But how, then, might entire clusters of stars form? If individual stars are formed by the gravitational collapse of a globule of gas like the ones found within larger interstellar nebulae, what might be the scenario for the formation of the hundreds or thousands of stars that constitute a cluster? Is it possible, for instance, that star clusters are formed from huge interstellar nebulae such as the Lagoon Nebula (p. 156), the Rosette Nebula (p. 157), and the Eagle Nebula (p. 158)? Are these giant luminous clouds of gas in fact star clusters in the process of forming?

These are interesting and potentially fruitful suggestions, but such hypotheses must pass numerous additional tests before they become viable theories concerning the formation of star clusters. For example, if large interstellar nebulae are the progenitors of clusters, we should expect to find substantial evidence for star-forming activity within these nebulae at this very moment.

In fact, we do find an abundance of such evidence. Within large interstellar nebulae we find numerous globule-like concentrations of gas and dust—precisely the kind of structures required for the birth of individual stars. (Approximately one percent of the mass of interstellar material comes in the form of small solid particles, usually referred to as "dust." It is this dust component in the more dense regions that obscures our view of parts of the glowing nebula, creating the dark structures evident in photographs.)

Furthermore, if stars are forming or have recently formed within a nebula, there should be some evidence of the heating action resulting from the gravitational collapse of globules and from the birth of thermonuclear-powered stars. Infant stars located deep within the nebula would probably be obscured from our view by the dusty environment, but if the interstellar dust component of the nebula is absorbing the visible light

NGC 6523—the "Lagoon" Nebula in Sagittarius. The dark structures visible in this photograph (and in the photographs on the next two pages) are relatively dense "globules" of gas and dust that block our view of the luminous material behind them. These structures are candidates for the formation of stars through the process of gravitational collapse. Palomar Observatory Photograph

generated by newborn stars, the warming action of that light-absorption process would lead the dust particles to reradiate that energy in the form of infrared radiation—the same phenomenon we experience as the "radiant heat" emitted from hot coals in a fireplace on a quiet winter's eve.

Are such infrared-bright "hot spots" observed? Indeed they are. Telescopes equipped to detect infrared radiation have revealed that a typical large nebula harbors dozens of localized heat centers just as our

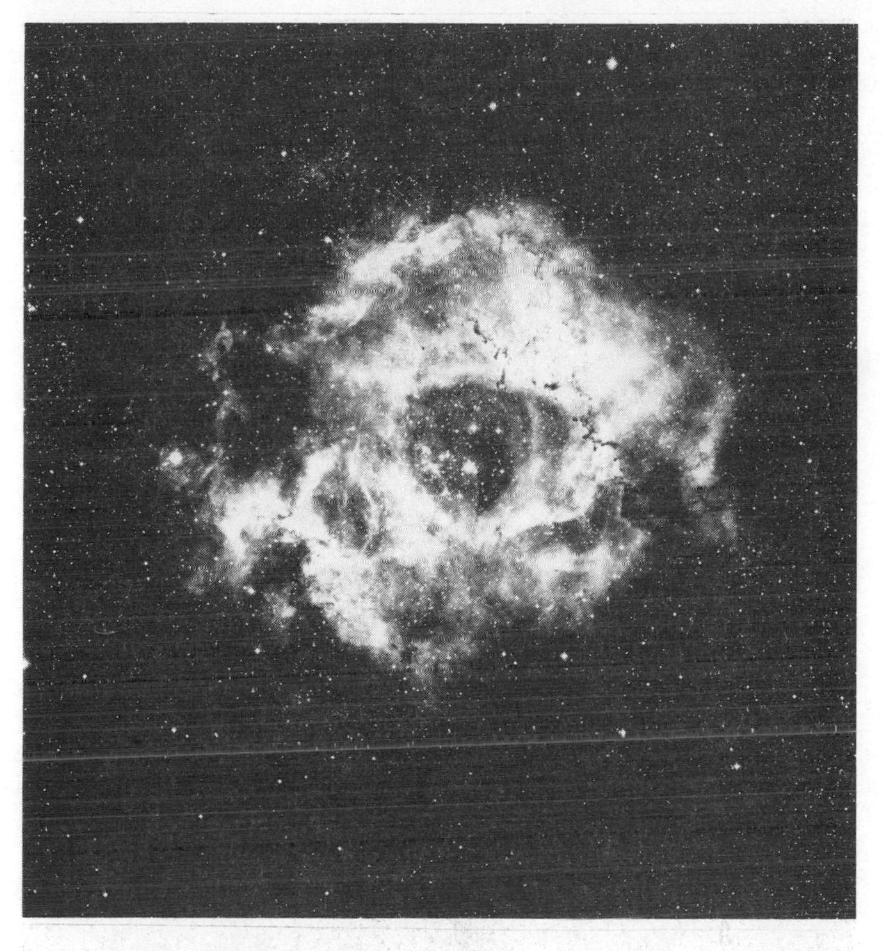

NGC 2237—the "Rosette" Nebula in Monoceros Palomar Observatory Photograph

model would predict. Our proposed scenario for star cluster formation thereby gains credibility.

But not all newborn stars would necessarily be obscured from view. We would expect that a few recently formed stars that are gravitationally held in association with the parent nebula would be visible to an outside observer. Our model would predict, therefore, that some newborn stars should be observed in the vicinity of stellar "maternity wards." Again, the observational evidence is plentiful. Associated with most large nebulae are a number of stars, among them two categories that readily reveal their youth.

Blue giant stars are massive main-sequence stars that are both very

NGC 6611—the "Eagle" Nebula in Serpens Palomar Observatory Photograph

luminous and very hot. Such stars are found almost exclusively in association with the kind of nebulae that we are here discussing. Why should this be the case? For the simple reason that these massive stars are short-lived and would therefore be found only near their place of birth. Blue giants, as a consequence of their high mass values, are gravitationally compressed to extremely high values of temperature and density, thereby leading them to consume their thermonuclear fuel rapidly. Such stars complete their main sequence phases relatively quickly, and on the H-R diagram they soon vacate their positions near the upper end of the main-sequence region. The amount of time a star spends in its main-sequence phase, called its *main-sequence lifetime,* would be only a few million years for a typical blue giant star. Thus, a blue giant is necessarily a relatively young (that is, having an age value small relative to the multibillion-year time span of cosmic history) and recently formed star. The observed association of blue giant stars with large interstellar nebulae is as natural as the association of newborn babies with a maternity ward.

T Tauri stars are less massive stars that have recently formed but have not yet stabilized into steady, long-lived, main-sequence stars. Readily recognizable because of their erratic variability and characteristic peculiarities in the wavelength spectrum of their light, T Tauri stars, like blue giant stars, are commonly found in conjunction with interstellar nebulae but nowhere else. Once again, the association of infant stars and stellar nurseries cries out to be noticed.

Clearly, then, the idea that clusters of stars form from massive interstellar nebulae is very strongly supported by the observational evidence. These nebulae have sufficient mass to make thousands of stars, concentrated "globules" of gas are a normal part of a nebula's structure, infrared "hot spots" give evidence of star-forming activity taking place within the nebulae, and young, recently formed blue giant and T Tauri stars are commonly found to be associated with star-forming nebulae. To propose, therefore, that a cluster of stars is formed from the material that had earlier composed a massive interstellar cloud is not merely a wild guess; it is a highly probable conclusion drawn from a wealth of coherently interrelated evidence. All of the known patterns for material behavior are fully incorporated into this theory for cluster formation, and a wide variety of empirically obtained information concerning the structure and behavior of nebulae is unified by this concept. Although any single piece of evidence might by itself be inconclusive, the extensive collection of consistently related phenomena provides the basis for a strong case in support of our theory.

One more question, however, needs to be addressed—the question of timetable. Over what span of time might this transformation from nebula to cluster take place? The evidence would seem to suggest that the majority of the star-forming activity takes place during a relatively short period of time—within a few million years, say.

Several specific physical mechanisms collectively promote this burst of star-forming activity. For example, when a large cloud of interstellar gas moves through a galactic spiral arm it is exposed to compressive forces sufficient to trigger the gravitational collapse of some globules of gas. And once star-forming activity begins, more is likely to be promoted. The presence of newborn stars, for instance, would promote a redistribution of gas within the nebula, and the supernova explosions of massive short-lived stars would send compressional shockwaves throughout the surrounding cloud. For a number of such reasons we are led to expect that the vast majority of stars in a particular cluster would be formed within a relatively brief span of time.

With the inclusion of these time scale considerations, our proposed scenario for cluster formation has become even more specific. Furthermore, we will demand that the subsequent development of a cluster must conform to the outline for stellar evolution briefly described in the previous chapter. All relevant factors, we insist, must be consistently incorporated into our theoretical model.

Testing the Model

We are now ready for the crucial test. Does our hypothesis, our theoretical model for the formative history of star clusters, provide us with a satisfactory means of understanding why the H-R diagrams of observed clusters should exhibit the specific peculiarities described at the outset of this discussion? As we soon shall see, it does indeed; and, as a welcome bonus, it also provides us with a means for determining the age of any star cluster.

Suppose, now, that a large cloud of interstellar gas were to undergo the transition from nebula to cluster in the manner of the scenario that we just developed. Further, suppose that we could inspect this hypothetical cluster shortly after the initial flurry of star-forming activity. How would the stars of a newborn cluster be distributed on an H-R diagram? According to standard theories of star formation, the infant stars would be zero-age main-sequence stars. (The "zero-age" label calls attention to the recentness of their birth.) And because of the diversity of conditions prevailing throughout the nebula, we should expect that among the newborn stars a wide variety of mass values would be found. Consequently, the full length of the main-sequence band should be observed—from the high-mass area in the upper left corner to the low-mass region in the lower right corner of the H-R diagram.

A couple of additional features might also be expected to characterize a newborn cluster. First, some of the original interstellar material might still be visible in the vicinity of a young cluster. Eventually, this gas and dust that did not go into star formation will be swept out of the cluster by the action of both starlight and stellar wind (an outflow of high-speed particles produced by most stars). Second, because low-mass globules collapse and contract more slowly than massive ones do, the stars toward the lower end of the main sequence might still be somewhat toward the right of their equilibrium position, as shown in Figure 2. (The transition from collapsing globule to zero-age main-sequence star can be traced out on the H-R diagram. As stars settle into their relatively stable positions along the main sequence, they ordinarily approach equilibrium from the low temperature area on the right side of the diagram.)

As a cluster matures, several changes should take place: leftover gas and dust should be flushed out of the area; the lower end of the main sequence should settle into its stable configuration; and the more massive stars should complete their main-sequence phases. The most massive stars, such as the blue giants found at the upper end of the main sequence, have the smallest values for main-sequence lifetime and consequently will be the first to complete that phase of their history and to vacate the main-sequence region on the H-R diagram. As the cluster ages, progressively less massive stars—those with correspondingly larger values of main-sequence lifetime—will come to the end of their main-sequence stage and trace out a path toward the red giant region on the diagram. If we could observe several billion years of cluster history on an H-R diagram, one of the more striking features of the observable change would be the "peeling away" of the upper end of the main sequence toward the right hand side of the diagram, as illustrated in Figure 3. As a cluster grows older, the point at which the "peel" (picture a banana being peeled) breaks away from the main sequence would move progressively down the length of the main sequence.

How does this predicted behavior compare with the H-R diagrams for actual clusters? Remarkably well! Recall our earlier description of those H-R diagrams: a portion of the upper end of the main-sequence region is unpopulated; and a string of stars extends from the turnoff point toward the red giant region. Both of these peculiar features are fully accounted for by our theoretical model based on a realistic combination of concepts drawn from the observed properties of interstellar nebulae, from the theory of star formation by gravitational collapse, and from the results of stellar evolution computations. *Our theory for cluster formation and development provides a systematic and consistent account for the H-R diagram peculiarities to which we had earlier drawn*

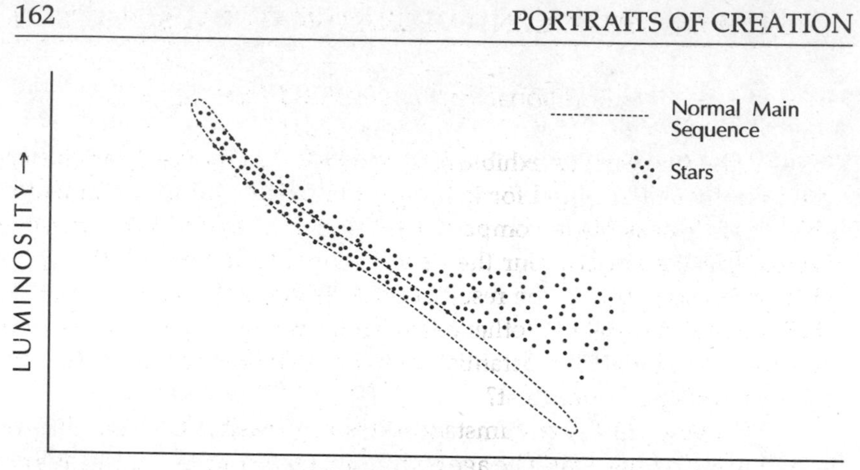

Figure 2. H-R Diagram of Stars in a Very Young Cluster. Low-mass stars, slow to come to equilibrium, have not yet reached their stable main-sequence positions along the lower portion of the main-sequence region.

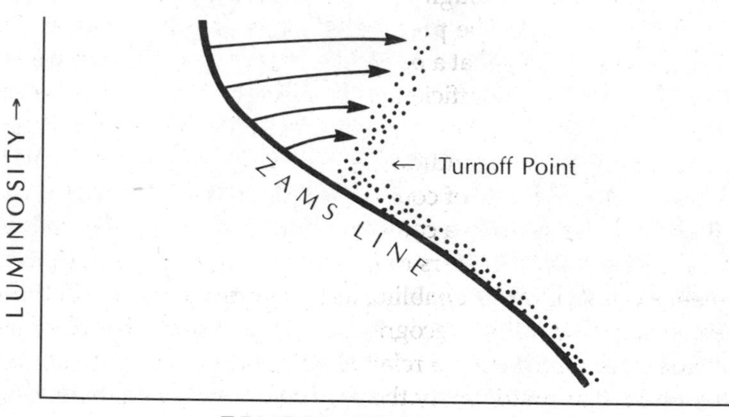

Figure 3. H-R Diagram of Stars in an Aging Cluster. The most massive stars have already completed their main-sequence phase and vacated their earlier positions along the upper portion of the main-sequence region.

attention. A model that was based on one set of considerations now functions to provide a remarkably coherent account of another set of data, thereby increasing our confidence not only in the credibility of this specific theory but also in the several components that contributed to this model for the formative history of star clusters.

Additional Applications and Tests

Now, if the peculiarities exhibited by the H-R diagrams of star clusters can be so well accounted for in terms of the historical development of clusters, is it possible to compute the duration of that history? In other words, can we employ our theoretical model to determine the *age* of clusters? And if so, do the results of this additional application of the theory fit coherently with other relevant chronological concepts already considered to be well substantiated? What happens when we put our theory to this additional test?

Under ordinary circumstances it is not possible to determine the age of an individual star. The age of the Sun, for example, can be inferred only from its presumed equality with the age of all of the other members of the solar system. And in our earlier reference to blue giant stars, the most that we could say about their age was that if they were still in their main-sequence phase, their age could not exceed the value of their main-sequence lifetime. Although this upper limit for the age of a star may be useful, it still lacks the precision that we might hope for. The problem is that when we look at a particular main-sequence star, we are generally unable to tell with sufficient precision whether that star is near the beginning, the middle, or the end of its main-sequence stage. Numerous uncertainties must be candidly recognized. (To overlook them would be both dishonest and, of course, without epistemic merit.)

But star clusters possess a combination of features that permit us to recognize that certain members of the cluster are at the end of their main-sequence phase, thereby enabling us to determine the age of those stars. The first step toward this recognition is to note that since all of the stars of a cluster formed during a relatively brief time interval, all stars of that cluster have approximately the *same age.* Next, we note that the stars that vacated the upper portion of the main sequence region above the turnoff point are stars that have already completed their main-sequence phase. Or, to express it in another useful way, the main-sequence lifetime values for these massive stars is *less than* the age of the cluster. On the other hand, the less massive stars below the turnoff point are still in their main-sequence phase because their main-sequence lifetime values are *greater than* the cluster's age. Stars located right at the turnoff point are a special "borderline" case. In the context of what we have just noted concerning stars above and below the turnoff point, we may safely conclude that *stars at the turnoff point on the cluster's H-R diagram are just completing their main-sequence stage;* they will be the next stars to vacate the main-sequence band and head toward the red giant region. How much time has elapsed since the formation of all stars in the cluster, including these? An amount of time equal to the value of the main-sequence lifetime of stars at the turnoff point. For these special stars—special because of their place on the H-R diagram, and special because their context allows us to determine the stage of their development—we are able to state that their age is *equal to* the value of their main-sequence lifetime.

The numerical value of this age can be determined with reasonable precision. The locations of stars along the main sequence are fixed by their mass values. Thus, from the location of the turnoff point we may compute the mass value for stars at that location along the main sequence. But the value of the main-sequence lifetime of a star is a known function of its mass value.[16] Putting all of these factors together, the location of the turnoff point on the H-R diagram can be readily translated into the value of the age of the cluster. Specifically, *the age of a star cluster is equal to the value of the main-sequence lifetime of stars located at the turnoff point on the H-R diagram for that cluster.*

Applying this procedure to both galactic and globular clusters leads to several noteworthy results. Globular clusters, located in the halo region of our galaxy, all have the same age value of approximately 13 billion years. This strongly suggests that all of the globular clusters formed at the same time that the galaxy itself was just forming. Galactic clusters, located in the gas-rich galactic disk, are found to have a wide variety of ages, from newborn to about 8 billion years.

Interestingly, the distribution of cluster ages appears to be closely correlated with the location of clusters within the different structural components of the galaxy. The absence of young clusters in the halo region, for instance, is consistent with the fact that the halo region is entirely lacking in the gas and dust from which clusters could form. On the other hand, the disk region, which includes the spiral arms, is rich with gas clouds—consistent with the presence there of numerous young clusters. (Very young clusters, by the way, display the two features predicted earlier: [1] the slow-forming stars along the lower end of the main sequence appear displaced toward the right of their eventual positions on the H-R diagram, and [2] very young clusters, such as the Pleiades, are still accompanied by significant amounts of gas and dust, presumably remnant materials from the original nebula.) In addition,

16. For brief discussions of this concept, see Shu, *The Physical Universe*, p. 146, and Howard J. Van Till, *The Fourth Day: What the Bible and the Heavens Are Telling Us about the Creation* (Grand Rapids: Eerdmans, 1986), p. 151.

the relative scarcity of galactic clusters having ages greater than a few billion years is consistent with the presence of several forces that tend to disperse the stars of these loosely structured clusters, thereby depleting the population of older galactic clusters. Although many galactic clusters may have formed early in galactic history, few of them would be expected to remain intact.

Many more details could be added to this discussion of star clusters, but what we have chosen to include is sufficient for our present purpose: to illustrate the manner in which a system of values functions in the scientific enterprise. Presuming that matters of *craft competence* and of *professional integrity* are not under question, we intend that this star cluster discussion function principally as an illustration of the typical way in which a system of *epistemic values* functions in the formulation and evaluation of a scientific theory. Recall, for example, how we insisted upon cognitive relevance—that an adequate theory relating the properties, behavior, and formative history of star clusters must account for the wealth of observational data concerning clusters. And wherever possible we encouraged the formulation of specific predictions that were amenable to empirical testing. From beginning to end we demanded that the theory be internally coherent and also that it be consistent with all other relevant phenomena—not only with all of the general laws of physics but also with other specific theories that were applicable to some aspect of cluster behavior. We sought a model that was as inclusive as possible, applicable to a diversity of phenomena, and able to unite a broad spectrum of other particular theories. And we valued the fertility of our theory, not only in the way it promoted further investigation of cluster properties but also in its function as an investigative tool concerning the formative history of the entire galaxy. As is the case for most substantive theories in contemporary natural science, the model for star cluster formation and development does not stand alone. It is consistently and intricately related, both qualitatively and quantitatively, to a mountain of empirical data and to a host of other well-substantiated theories. The goal of such theorizing is neither more nor less than to apprehend the intrinsic intelligibility of the physical world in which we live.

For what purpose do we seek this knowledge? How shall we employ this knowledge? What is the larger context that provides for us the meaning and significance of this knowledge? These important questions take us far outside of the limited domain of the natural sciences, and in our quest for answers we shall have to look to sources beyond the object of scientific investigation. The scientifically accessible portion of the picture of reality is often fascinating, but ultimately it is incomplete.

6. A CRITIQUE OF THE CREATION SCIENCE MOVEMENT

ROBERT E. SNOW

During the past quarter century a vigorous movement known as "creation science" (or "scientific creationism") has emerged, primarily out of the fundamentalist portion of the North American Christian community. Its strong defense of a recent special creationist picture of God's creative work is built primarily on a scientific appeal to empirical data that have been selected and reinterpreted to conform to a particular reading of the Bible. In light of the widely accepted scientific world picture, the central claim of creation science is truly astounding: the entire universe, when investigated in accordance with the canons of the creationist program, shows convincing scientific evidence for its having been brought into existence recently in the fully functioning and structured form that it has today.

How well do such extraordinary claims hold up under close scrutiny? What are the functioning dynamics of this movement in relationship to the worldviews (both Christian and non-Christian) and world pictures (both scientific and nonscientific) found in contemporary culture? Robert E. Snow provides us in this chapter with an in-depth analysis and an insightful critique of this eccentric and puzzling perspective.

INTRODUCTORY COMMENTS

THE EMERGENCE OF THE CREATION SCIENCE movement during the past two decades has been an extraordinary development within the arena of popular science and popular religion.[1] Whether it is also the harbinger of a radically reformed "Science of Tomorrow" is the major question this chapter will address. If we listened only to the claims of leading creationists, there is no doubt what the answer would be. An interview with Thomas G. Barnes published in *Christianity Today* provides a typical example:

> The clarity and reasonableness of [creation-science] postulates, the great scope of the observational data, and the strict adherence to fundamental laws of physics included in the evidences for a young earth are superior to the presumed evidences for an old earth.[2]

Henry M. Morris echoes the assertion:

> The real facts of science, as distinguished from various interpretations

1. In this chapter we use the term *creationist* to denote someone who advocates the development of a science consistent with the view that creation was completed in six twenty-four-hour days, that the Earth and the universe are young (approximately 10,000 years in age), and that most of the Earth's geological features are attributable to the action of a worldwide flood. In an article in the *Christian Century* (24 April 1985, p. 411), Conrad Hyers claimed there were in print "more than 350 books challenging evolutionary science and advocating a 'creation science' " approach. In Appendix C of his *History of Modern Creationism* (San Diego: Master Book Publishers, 1984), Henry M. Morris provides a list of 109 creationist organizations "whose *primary* purpose is apparently to research, promote, teach, and/or disseminate information in support of scientific and/or Biblical creationism" (pp. 341-47). The list includes twenty-two national U.S. organizations, fifty-four state and local organizations, and thirty-three foreign organizations representing sixteen different countries. The great majority of these organizations clearly are devoted to grass-roots dissemination of creationist ideas rather than to the doing of creation research. The Institute for Creation Research is the most important creationist organization devoted to developing the scientific side of creationism, but even its staff devote much of their energies to popularization. An informal account of staff activities from 1972 to 1984 mentions "lectures or messages in probably 1000 churches and in at least 500 colleges and universities, plus many other types of meetings and audiences, several hundred guest appearances on radio and television programs, a 15 minute radio program produced every week for the entire period, a monthly newsletter, and more than 150 creation/evolution campus debates" (*History of Modern Creationism*, pp. 249, 255-66). The *History of Modern Creationism* is the best single source for information about modern creationism. Partly autobiographical, it is an insider's informal history.

2. "Gish and Barnes Field Questions," *Christianity Today*, 8 October 1982, p. 36. A professor emeritus and former physics department chairman at the El Paso campus of the University of Texas, Barnes was a member of the original steering committee that guided the development of the Creation Research Society. After retirement from El Paso, he served as dean of the graduate school of the Institute for Creation Research.

> imposed on those facts, all point to the recent special creation of all things, not long ages of evolutionary uniformitarianism.[3]

In contrast, such claims are usually dismissed out of hand by opponents. For example, Kenneth R. Miller claims that

> The creationists, realizing that the enormous weight of scientific evidence is stacked in favor of evolution . . . [have] concentrated instead on political lobbying and on taking [their] case to a fair-minded electorate. . . . The reason for this strategy is overwhelmingly apparent: no scientific case can be made for the theories they advance. Therefore, rather than present these ideas to an audience of specialists (geologists, geneticists, molecular biologists, biochemists, and paleontologists) who would at once point out the factual contradiction in their ideas, they have chosen instead the general audience of the interested public.[4]

What are we to make of claims that conflict so dramatically? Even allowing for the rhetorical escalation that inevitably accompanies public controversy, we seem to be confronted by radically different visions of the world.[5] In one major respect the present situation with regard to the claims of creationists differs remarkably from that of the twenties, thirties, forties, and fifties. During those decades the principal representatives of creationism, such as George McCready Price and Harry Rimmer, were self-taught amateurs without formal scientific credentials.[6] Today the leading creationists have advanced degrees in the basic

3. Morris, *The Biblical Basis for Modern Science* (Grand Rapids: Baker Book, 1984), pp. 125-26. Morris is the single most important figure of the modern creationist movement and has been involved in virtually all of its major developments. His writing style is very clear, and he is evidently a skilled and energetic administrator.

4. Miller, "Scientific Creationism versus Evolution," in *Science and Creationism*, ed. Ashley Montagu (New York: Oxford University Press, 1984), pp. 22-23.

5. Often enough it will happen that in matters of scientific controversy, what one side holds to be "facts" will be characterized by the other side as "myths." For an example of this, see the discussion of the controversy over safe levels of public exposure to lead in David Robbins and Ron Johnston, "The Role of Cognitive and Occupational Differentiation in Scientific Controversies," *Social Studies of Science* 6 (1976): 357-58.

My use of the terms *world* and *vision* is deliberate and is intended to reflect the framework of analysis suggested by James H. Olthuis, "On Worldviews," *Christian Scholar's Review* 14 (1985): 153-64.

6. According to Ronald L. Numbers, "Price attended an Adventist college in Michigan for two years and later completed a teacher-training course at the provincial normal school in his native New Brunswick" ("The Creationists," in *God and Nature: Historical Essays on the Encounter between Christianity and Science*, ed. David C. Lindberg and Ronald L. Numbers [Berkeley and Los Angeles: University of California Press, 1986], p. 400). Henry Morris adds that Price received a B.A. from Loma Linda College in 1912. By the date of his degree he "had already been teaching there in such subjects as Latin, Greek, Chemistry, and Physics, as well as general refresher courses in all fields for students in

or applied sciences and are supported by a well-established set of organizations. These include, most notably, a professional society (the Creation Research Society, established in 1963), a professional journal (the *Creation Research Society Quarterly*, established in 1964), a research center staffed by professionally trained scientists (the Institute for Creation Research, established in 1972), and a publishing house (Creation-Life Publishers, established in 1974). By 1989 voting membership in the Creation Research Society, requiring at least a masters degree in the natural or applied sciences, had grown to 635.[7] The broad public credibility gained by creationists since 1970 is closely related to their claim to professional scientific credentials. At first glance both creation scientists and the scientists opposing them seem to have appropriate credentials for making the claims they make. But they disagree, and the disagreement is vehement and bitter.[8]

Creation science is a remarkable movement. It is remarkable both for the widespread public support it has gathered and for the increasingly extreme stance creationist leaders have taken with respect to the broad spectrum of scientific theory. In 1972 Henry Morris wrote, "real creation necessarily involves creation of 'apparent age.' Whatever is truly created—that is, called instantly into existence out of nothing—must certainly look as though it had been there prior to its creation."[9] More recently, however, he has claimed that the earth not only *is* young, but also that it *looks* young when viewed through the prism of science: "There are many times more geological processes and systems that yield a young age for the earth than the handful of radiometric methods that can be forced (through an extreme application of uniformitarianism) to yield an old age."[10]

---

pre-medicine and pre-nursing" (*History of Modern Creationism*, p. 81). According to Morris, Rimmers's formal education was very limited and heterogeneous (a year each at Hahneman College of Homeopathic Medicine, San Francisco Bible College, Whittier College, and the Bible Institute of Los Angeles [*History of Modern Creationism*, p. 80]).

7. *Creation Reasearch Society Quarterly* 26 (1989): 101. The total 1989 membership is reported to be 1,764, of which 36 percent are voting members.

8. Obviously a substantial element in any final assessment of the creation science movement must be an appraisal of their claims as interpreters of the Bible. Such an assessment will not be attempted in this chapter, although it forms part of the broader context within which the claims of creation scientists should be evaluated.

9. Morris, *The Remarkable Birth of Planet Earth* (Minneapolis: Dimension Books, 1972), p. 62.

10. Morris, "Recent Creation Is a Vital Doctrine," *Impact* no. 132, Institute for Creation Research, June 1984. For a detailed examination of the evidence that creationists use to support their often repeated claim that there are many "geological processes and systems that yield a young age for the earth," see the case study "Timeless Tales from the

This claim has dramatically widened the front across which the creationists are committed to challenging contemporary science. It entails at a minimum the rejection of biological evolution and the main body of historical geology, plus most of quantum physics and astrophysics. It is not surprising that creationist positions are usually given short shrift by the scientific community. In matters of theory and in details of interpretation, they are at odds with many of the most prestigious aspects of contemporary science.[11] If sustained, they would revolutionize our understanding of the created world.

Salty Sea" in Howard J. Van Till, Davis A. Young, and Clarence Menninga, *Science Held Hostage: What's Wrong with Creation-Science AND Evolutionism* (Downers Grove, IL: Inter-Varsity Press, 1988), pp. 83-91.

11. For many creationists the universe-actually-looks-young claim functions as a methodological principle in ways analogous to the uniformitarianism they often oppose: it serves to endorse many arguments that appear hopelessly ad hoc to someone convinced that uniformitarian strategies provide the most fundamental guidelines for scientific argument. The creationist tendency to employ ad hoc arguments is also encouraged by the conceptual uncertainties of a young universe. The considerable coherence of our contemporary understanding of the universe is intimately bound up with the concept of great age. We all (creationists and noncreationists alike) have great difficulty imagining the structure of theory, law, and principle that will yield an equivalent coherence for a young universe. Creationist claims often raise hackles because in our contemporary world there is a fundamental strangeness inherent in the universe-actually-looks-young principle. This basic "strangeness" makes it extremely important that their case be made with great care. Actually, that seems to be the exception rather than the rule.

In *The Biblical Basis for Modern Science,* Henry M. Morris provides many illustrations of the "strangeness" effect. The following is taken from his discussion of the solar system:

> From the point of view of both God and man (by whom and for whom, respectively, the Word of God was revealed), of all the physical bodies in the universe, the earth is most important, then the sun and moon, then the stars. Therefore, the earth was created first (Gen. 1:1), then the two great lights to rule the day and night (Gen. 1:16a), and finally "the stars also." (Gen. 1:16b)
>
> This is the reverse of both the importance and chronological order imagined by evolutionists, according to whom the universe evolved first, then its galaxies of stars, and finally the solar system, with the earth and moon somehow spinning off from the sun in the process. Although it is clearly impossible to prove scientifically which of these two sequences is correct, the Biblical order is far more logical. The earth is the most complex body in the physical universe, so far as known. The moon is much less complex than the earth, and the sun (consisting mostly of hydrogen and helium) is very much less complex than the moon. The various stars (since there is no evidence that any of them support planetary systems) are probably even less complex than the sun. Intrinsic value, of course, is measured by organized complexity—by "information"—rather than mere size. . . . This Biblical order—earth, sun, moon, stars—will seem shocking to evolutionists, of course, but there is nothing either impossible or illogical about it in the context of God's creative power and purpose. (Pp. 161-62)

THE SHRINKING SUN: PROBLEMS WITH THE EXERCISE OF PROFESSIONAL INTEGRITY

How can we begin to develop a vantage point from which to assess the credibility of statements made by creation scientists such as Barnes and Morris? One way is to take a close look at the evidence they offer in support of their position. For example, in recent years creationist leaders have often referred to evidence that the Sun is shrinking as providing important support for their position. To understand and assess such a claim it is important to trace the relationship between this particular creationist claim and the development of the "shrinking Sun" discussion among the group of practicing scientists who have the craft competence to carry out the appropriate research.[12] What follows is a case study in two parts. The first part is a short history of the shrinking Sun controversy among professional astronomers. It is a typical example of the way in which minor controversies are handled within a community of practicing scientists. Such controversies are a standard feature of scientific life. The second part of the case study examines the treatment of the controversy by creationist writers and leaders.

SCIENCE PROCEEDS not only by the exploration of unknown territory but also through the critical revision of previously accepted results. Fresh insights generated by novel theoretical advances, surprising experimental results, and even the careful reanalysis of long-available data may serve to provide the necessary stimulus for such critical revision. As a consequence, the professional scientific literature is pervaded by publications that raise questions about received views. If the question raised is interesting enough, the common result is a small flurry of publications as researchers probe both the assumptions behind the novel suggestion and its implications for future research. Depending on the results of the research, the claim that initially attracted attention may become firmly established as part of a new orthodoxy, incorporated into the research tradition in a highly modified form, or simply dismissed as a mistaken enterprise. In any event, when the burst of investigation in the matter has run its course, the researchers directly affected turn to new issues within their chosen field. And whatever the result, the public legacy of the affair is a set of published reports partially recording another micro-episode in the continuing process of critical assessment and appraisal

12. For a discussion of "craft competence" see our chap. 5, pp. 138-39.

that is the life blood of professional science. A more subtle consequence of the flurry of activity is reflected in changes in the unwritten lore of the field concerning such issues as experimental design, data reduction, data evaluation, and appropriate patterns of inference.[13] Such episodes provide an illustration of the extent to which science is a collaborative social enterprise of researchers, journals, editors, reviewers, conferences, and even reporters—all linked together by formal and informal networks of communication.

In 1979 a portion of the solar physics community attending the June meeting of the American Astronomical Society was startled to hear the highly respected astronomer J. A. Eddy, working with A. A. Boornazian, report that solar meridian transit data collected at the Greenwich Observatory from 1836-1953 showed the Sun's diameter was shrinking at the rate of five feet/hour or 0.1 percent per century.[14] The report, if substantiated, had significant theoretical consequences in a number of areas of solar studies and immediately attracted the attention of researchers. Because of its potential significance, a popular account of the research, including interviews with a number of leading solar physicists and astronomers, was published in the September issue of *Physics Today*.[15] Everyone interviewed judged that the extremely high rate of shrinkage reflected some kind of oscillatory phenomenon occurring in the Sun's outer layers. The article also included reports of several lines of research underway, or soon to be started, that were designed to provide a cross-check upon the accuracy of the original report. In the next four years a number of papers were published in the professional scientific literature.[16] In the process serious questions were raised about both the intrinsic accuracy of the original data and the methods used by Eddy and Boornazian to correct for atmospheric and instrumental

13. By now there is a large, controversial, and rapidly growing literature discussing the relationship between the social and cognitive aspects of scientific research. Early and still helpful contributions to understanding this aspect of science were Michael Polanyi, *Personal Knowledge* (London: Routledge & Kegan Paul, 1958); Thomas Kuhn, *The Structure of Scientific Revolutions* (Chicago: University of Chicago Press, 1962); and John Ziman, *Public Knowledge* (London: Cambridge University Press, 1968). For an expanded discussion of "tacit" knowledge, see our chapter 5, pp. 146-47.

14. Eddy and Boornazian, "Secular Decrease in the Solar Diameter, 1836-1953," *Bulletin of the American Astronomical Society* 11 (1979): 437. The term *secular* is used here in the technical sense "of or relating to a long term of indefinite duration."

15. G. B. Lubkin, "Analyses of Historical Data Suggest Sun Is Shrinking," *Physics Today* 32 (1979): 17.

16. For a detailed account of the various journal articles that constitute the scientific literature of the "shrinking Sun" controversy, see the case study by Howard Van Till in *Science Held Hostage*, pp. 42-65.

changes. In addition, estimates of the solar diameter derived from other kinds of observations showed little, if any, long-term change in the Sun's diameter. By 1983 a consensus was emerging that good evidence existed for "an (approximately) 80-year cyclic variation."[17] By 1984 Eddy (with C. Frohlich) was reporting that data from 1967-1983 indicated a mean *increase* of eight feet/hour in the solar diameter—a result consistent with an eighty-year cycle—but remarkably different from Eddy's position five years earlier.[18]

While the original claims of Eddy and Boornazian were not sustained, the solar physics community emerged from the exchanges generated by their initial paper with a better understanding of the strengths and limitations of important data sets gathered over long periods at major observatories. More importantly, close analysis of that data revealed evidence for a previously unrecognized cyclical variation in the Sun's diameter. Eddy and his collaborators, because they continued to participate in the process of public criticism and analysis, maintained their status as valued members of the professional community. The episode provides a good illustration of the healthy operation of a professional community.

At the same time a parallel, but quite different, scenario was unfolding within the creationist community. Because many creationists believe that any evidence against evolution is evidence for their position, they are constantly looking for results that are anomalous within an evolutionary framework, for theoretical inconsistencies, and for disputes within evolutionary disciplines that can be turned to their advantage. By itself, this is perfectly legitimate. The critical questions concern how the anomalies, inconsistencies, and disputes are used by creationists.

In the shrinking Sun episode, creationist writers responded quickly to the initial June 1979 report by Eddy and Boornazian. In September, October, and November of 1979 editors of the *Creation Research Society Quarterly* (hereafter referred to as *CRSQ*) received two manuscripts from Hilton Hinderliter and one from Paul M. Steidl that were published the next year in the June and December issues of the *CRSQ*.[19] In April 1980 a "shrinking Sun" article by Russell Akridge was

17. J. H. Parkinson, "New Measurements of the Solar Diameter," *Nature* 304 (1983): 518.

18. Frohlich and Eddy, "Observed Relation between Solar Luminosity and Radius," paper presented at an international conference sponsored by the Committee on Space Research, July 1984, in Graz, Austria.

19. Hinderliter, "The Shrinking Sun: A Creationist's Prediction, Its Verification, and the Resulting Implications for Theories of Origins," *CRSQ* 17 (1980): 57; and "The

published in the more popular *Impact* series, which is distributed free of charge by the Institute for Creation Research to a mailing list of 75,000.[20] Hinderliter and Akridge accepted uncritically the Eddy and Boornazian result and then asserted that it rules out the standard view that the Sun is powered by nuclear fusion. If the Sun were shrinking as fast as Eddy and Boornazian claim, then gravitational collapse by itself would supply more than enough energy to power the Sun. This is important for creationists, because a Sun heated only by the energies of gravitational collapse cannot be old enough to allow evolution to have taken place. The Steidl article is more careful, but in a qualified way it comes to the same conclusion.

The Hinderliter, Steidl, and Akridge articles represent the first phase of the creationist response to the shrinking Sun phenomena. They reject the suggestions of astrophysicists that the proposed shrinkage is cyclical and limited to the solar envelope as mere ploys designed to save the evolutionary time scale. When they note the lower estimates of Sofia, O'Keefe, Lesh, and Endal (published about the same time the Eddy and Boornazian paper was read), it is only to observe that they also are compatible with the gravitational collapse theory of solar energy.[21] This is a fair claim, and while the rhetoric of Hinderliter and Akridge is overblown, they are still in touch with the directly relevant professional literature.[22]

The second phase of the creationist response is represented by a single letter from Steidl to the editor of the *CRSQ*. Received in August 1980, but not published until the following March, it warns readers of the *Quarterly* that the Eddy and Boornazian claim "has met with much skepticism, and the timing of transits of Mercury over the centuries seems to indicate that no shrinkage is taking place."[23] His warning was appropriate, and it accurately reflected the changing state of scientific opinion and evidence. Steidl's closing comment that "only time will tell

---

Inconsistent Sun: How Has It Been Behaving, and What Might It Do Next?" *CRSQ* 17 (1980): 143; and Paul M. Steidl, "Solar Neutrinos and a Young Sun," *CRSQ* 17 (1980): 63.

20. Russell Akridge, "The Sun Is Shrinking," *Impact* no. 82, Institute for Creation Research, April 1980. Akridge is a physicist at Oral Roberts University.

21. S. Sofia, J. O'Keefe, J. R. Lesh, and A. S. Endal, "Solar Constant: Constraints on Possible Variations Derived from Solar Diameter Measurements," *Science* 204 (1979): 1306.

22. Hinderliter writes, "it is clear that we have witnessed a *major scientific defeat for evolutionism*" ("The Shrinking Sun," p. 59; italics his). Akridge closes his essay with the claim that "The changes detected in the sun call into question the accepted thermonuclear fusion energy source for the sun. This, in turn, questions the entire theoretical structure upon which the evolutionary theory of astrophysics is built."

23. Steidl, "Recent Developments about Solar Neutrinos," *CRSQ* 17 (1981): 233.

what the ultimate outcome will be and its importance to creationism" apparently went unheeded.[24]

The third phase of the creationist response is one in which creationist writers continue to echo the tone and substance of the earlier Hinderliter and Akridge comments. A June 1981 *CRSQ* article by James Hanson mentions—and dismisses without serious engagement—the literature that concerned Steidl.[25] Thomas Barnes included the Akridge version of the shrinking Sun as one of his six "evidences" for a recent creation when he was interviewed by *Christianity Today* in October 1982. In 1983 the *Origins Film Series Handbook* used a diagram from the Akridge article as it repeated his claims. In 1984 the pattern was repeated by Henry Morris: "Careful measurements in recent years have supported the collapse theory by showing that the sun's diameter does, indeed, appear to be shrinking. But this in turn would mean that the sun could not possibly be billions of years old!"[26] Completely detached from the scientific literature that five years earlier had suggested the Sun might be shrinking, Morris repeats a claim that had become a standard element in the creationist arsenal. Unfortunately, by 1984 it was legendary rather than scientific in character.

A closer analysis of the "shrinking Sun" episode would illustrate a number of serious faults, including (1) unwarranted extrapolation, (2) exclusion of relevant data, (3) failure to consider relevant processes or events, (4) failure to correct items 1-3 even after learning of the problem, (5) loss of contact with the professional scientific literature, and (6) dependence on secondary sources. Each of these faults reflects a significant failure to maintain appropriate standards of professional integrity. And unfortunately the "shrinking Sun" episode is not an isolated evidence of such failures.[27] The creationist system of publication does not function in a way that encourages the critical self-evaluation that is necessary for the good health of any professional field of science. Indeed, it often seems to provide a protected haven for poorly grounded creationist claims. In one sense, such a state of affairs should not be surprising. In their effort to develop a radically different kind of science,

24. Steidl's warning and its implications are not taken up in any of the creationist literature I have examined.

25. Hanson, "The Sun's Luminosity and Age," *CRSQ* 18 (1981): 27.

26. Morris, *The Biblical Basis of Modern Science,* p. 164.

27. For a series of extended case studies documenting the occurrence of similar problems in creationist arguments based on the amount of dust on the Moon, the concentration of various salts in the oceans, and the geology of the Grand Canyon, see Van Till, Young, and Menninga, *Science Held Hostage,* pp. 67-124.

creationists do need a "protected environment" in which ideas that deviate from conventional professional science can be tried out and polished. Unfortunately, as the "shrinking Sun" episode illustrates, reliance on protected environments can lead to very serious problems. The creationists' claim that they are seeking to develop a new science should be taken seriously, but it is very important—particularly for nonscientists attracted by the claims of the creation-science movement—to appreciate the difficulties inherent in their efforts and to refrain from uncritically accepting their work. As a further step toward understanding the tensions and dilemmas facing creationists who wish to do their work with integrity, it is useful to consider what is implied if we think of creationists as "sectarian" scientists.

SOME COMMENTS ABOUT SECTARIANISM AND CREATION SCIENCE

By definition, Christian sectarian movements exhibit one or more of the following attributes:

1. They add to Scripture.
2. They reject a significant portion of Scripture.
3. They elevate one theological principle over all others and give it an inappropriate control over their theological system.

The sectarianism of creation science gives evidence of embodying the third attribute. Its proponents are scientific sectarians in the strong sense that they add to the accepted set of values that scientists use in their appraisal of theories. And the value they add is not just another factor to be considered in theory appraisal; it is elevated to a controlling status. In brief, their controlling epistemic value is that all theoretical explanations must be in accord with their understanding of the character and chronology of creation. While many aspects of creation science are objectionable to the larger community of science, the most fundamental sticking point is the insistence by creationists that all theory should be evaluated in terms of their ruling epistemic value.

A leading student of sectarian groups, Thomas O'Dea, has suggested that, given appropriate conditions, the basic pattern of sectarianism can be found outside of religious settings. As the following quotations indicate, many of his comments about religious sects also apply to the creation science movement.

> The sect as a sociological ideal is . . . to be understood as the embodiment and expression of rejection of some significant aspect of secular life.[28]
>
> [The creation science movement arose in explicit rejection of a major defining element of modern secular society—the evolutionary world picture.]
>
> It is a voluntary society of strict believers who live apart from the world in some way. (P. 131)
>
> [Through the elaboration of their institutional network, creation scientists are increasingly able to maintain a professional life that is insulated from the professional setting of the broader scientific community.]
>
> The sect expresses defiance of the world . . . [and] a greater or lesser rejection of the legitimacy of the demands of the secular sphere. (P. 131)
>
> [Creation scientists defy the claims of conventional science. By and large, Henry Morris speaks for the whole movement when he claims that "all the *real facts of science* support . . . special creation of all things in six literal days and a world wide cataclysmic destruction by the flood."[29]]
>
> A correspondence and interpenetration of religious and social interests has often been found associated with the origin and formation of sects. (P. 132)
>
> [While the interpenetration of social interests and the religious and scientific concerns of the movement is mentioned only in passing here, it has played a vital role in the rise of creationism.[30]]

28. O'Dea, "Sects and Cults," in *International Encyclopedia of the Social Sciences,* vol. 14, ed. David L. Sills (New York: Macmillan, 1968), p. 132. Subsequent references to this article will be made parenthetically in the text.

29. Morris, *The Biblical Basis for Modern Science,* pp. 125-26.

30. In his *History of Modern Creationism,* Morris notes that as a result of his study of Scripture in the context of his daily interaction with "the unbelieving intellectual world . . . it [became] crystal clear to me that the foundation of false teaching in every discipline of study, and therefore of ungodly practice in all areas of life, was evolutionism" (p. 223). This perspective provided the basis for the close cooperation with Tim LaHaye that led to the founding of Christian Heritage College and the Institute for Creation Research. The college opened in the fall of 1970 under the sponsorship of Scott Memorial Baptist Church, which LaHaye pastored. Morris served as the first academic dean of the college and LaHaye was the president from its founding until 1978. Since 1970 LaHaye has risen to national prominence as a leader of the Moral Majority and is, as Morris puts it, "a national leader in the battle against the humanistic moral influences subverting our nation" (p. 225). To all appearances, Morris is fully sympathetic with the political agenda of the Moral Majority and views creation science as one of the chief weapons in the national and international battle for true belief and right moral action.

For a well-informed interpretation of the ways in which rival religious and social

At a minimum, creation science *mimics* many of the religious and sociological aspects of a sect.

One final observation by O'Dea should be considered because it highlights the most problematic aspects of trying to be a sectarian scientist:

> Some sects are successful in maintaining themselves over a long period of time in an established condition of opposition or at least nonacceptance, with respect to secular society and its values. . . . They succeed in keeping their membership apart from genuine, intimate, nonsectarian social participation. (P. 131)

What counts as a long period of time? As of 1990 the *CRSQ* is in its twenty-sixth year of publication and the Institute for Creation Research (the first creationist organization with a full-time staff devoted to research, writing, and teaching) is in its eighteenth year. In the professional life of a scientist these are relatively long periods of time. The *CRSQ* and the Institute are the two most important professional organizations of creation science, and they certainly function to keep their members "apart from genuine, intimate, nonsectarian social participation" in the life of the broader scientific community. They provide the only "professional" setting in which creation scientists can publicly exercise their special controlling epistemic value without incurring severe sanctions.

But what is wrong with that? Surely creation scientists need a professional setting in which they can be free to develop their ideas. The problem is not the existence of the *CRSQ* or the Institute (they are necessary for the development of the creationist research program) but the extent to which they separate creation scientists from the normal practice of science so that they lose touch with the full range of the several categories of knowledge and value that are essential to its competent practice. As we have seen, there is substantial evidence that the very professional organs that encourage the social flourishing of creation science are providing an environment that militates against the achievement of adequate scientific results.

The fundamental problem is that the nitty-gritty of scientific competence and integrity is embedded in the craft knowledge of each subdiscipline and field of research. Appropriate craft knowledge can be learned and maintained only by continued participation in an active

---

meanings have been expressed in controversies about the age of the Earth, see Martin J. S. Rudwick, "The Shape and Meaning of Earth History," in *God and Nature,* pp. 296-321.

community of researchers. Furthermore, continuous participation in the life of an established research community is an essential safeguard against the unrecognized erosion of appropriate standards of craft knowledge and professional integrity. While much can be learned from reading technical reports, there is no substitute for the insight gained from working in an active research program that includes participation in the continuous round of publication and criticism in appropriate professional journals. To remain a competent scientist, it is necessary to continue to pay the proper professional dues. Creation scientists cannot expect to critique competently work they dislike, nor can they expect to develop their own alternative science, without continued participation in the working craft life of the existing research specialties of professional science. And that is hard to do while maintaining an appropriate sectarian separateness. This is the Catch-22 of competent creation science. When your close associates tell you that the "conflicting evidence" doesn't seriously conflict or doesn't *really* even exist (recall Morris's statement that all the "real facts of science" support special creation), it is hard to keep a high critical stance. This is creation science's weakest feature. Critics return again and again to the cavalier manner with which creation scientists often treat evidence that seems to conflict directly with their claims.

A CENTRAL PROBLEM FOR CREATION SCIENCE

Increasingly creationists have underscored their claim that the Earth, solar system, and universe not only *are* young but also *look* young from a scientific perspective. This has put additional stress on the importance of developing a plausible account of science that will allow them to discount the "historical" claims of geology, astronomy, and biology. Indeed, this is the fundamental problem for a creationist philosophy of science. The emphasis by Barnes and Morris on the "real facts of science" and the "great scope of observational data" supporting creation science, in contrast to "presumed evidences" and "interpretations imposed" upon those "real facts" by secular scientists, suggests that they believe the problem has been solved.

Among creationists, John N. Moore has devoted the most attention to working out a philosophy of science designed to discount the historical sciences. In his book *How to Teach Origins (without ACLU Interference),* it is the central focus of a forty-six-page discussion of scientific method.[31] Creationists typically emphasize the empirical and factual aspects of science, and Moore is no exception to the general pattern.[32] But he also recognizes that facts by themselves are worth very little. They become significant, he says, when the relationships between seemingly "diverse and apparently isolated facts" are brought into focus by a scientific theory.[33] Moore acknowledges that the explanatory power of theories often depends on their use of concepts such as atom, gene, or electron (he calls these "imaginary entities") that cannot be directly observed. At the same time he wants to be able to draw a line between theories that creationists are willing to accept (e.g., theories of the atom or the gene) and theories they are not willing to accept (e.g., stellar evolution and plate tectonics—theories that imply that the Earth or the universe is old).

Moore attempts to solve the problem by definition. The phrase "natural objects and/or events" appears repeatedly in his text. For example, "All proper scientific theories are based upon empirical aspects of scientific activity associated with *natural objects and/or events;* and, in turn, scientific theories are fruitful of further scientific observation and experimentation with respect to *natural objects and/or events.*"[34] This seems rather innocuous until some unusual restrictions on the term *natural* are factored in. According to Moore, "The term 'natural' def-

31. John N. Moore has long been a significant contributor to the creation science movement. He was a member of the original Board of Directors of the Creation Research Society and the first managing editor of the *Creation Research Society Quarterly* (a position he held for almost twenty years). An important creationist writer, Moore was the coeditor of *Biology: A Search for Order in Complexity* (Grand Rapids: Zondervan, 1974). His most ambitious book, *How to Teach Origins (without ACLU Interference)* (Milford, MI: Mott Media, 1983) contains a lengthy chapter entitled "What Is Science?" (pp. 55-100). Moore's interest in philosophy of science goes back at least to the early 1970s, when an article of his entitled "Evolution, Creation, and the Scientific Method" appeared in *American Biology Teacher* (January 1973, pp. 23-26).

32. In this respect creationists reflect their heavy debt to nineteenth-century Baconianism. Like their nineteenth-century predecessors, they pay little attention to the ways in which "facts" bear the impress of theories. Moore is typical when he claims that "the postulates of a *scientific* theory are always based upon observations of relevant natural objects and/or events that scientists have noted *before* they stated theoretical postulates" (*How To Teach Origins*, p. 81).

The debt of creationism to the nineteenth century is discussed by George M. Marsden in "Understanding Fundamentalist Views of Science," in *Science and Creationism*, ed. Ashley Montagu (New York: Oxford University Press, 1984), pp. 95-114. For a standard account of nineteenth-century Baconianism, see Theodore Dwight Bozeman, *Protestants in an Age of Science: The Baconian Ideal and Antebellum American Religious Thought* (Chapel Hill, NC: University of North Carolina Press, 1977).

33. Moore, *How to Teach Origins*, p. 79.

34. Moore, *How to Teach Origins*, p. 83; italics added.

initely refers to things and processes that actually exist and really occur in the physical environment, and are directly or indirectly accessible to natural scientists."[35] He wants to restrict the use of *natural* to events and/or processes that occur "during the life experience of some human beings [and] are either directly or indirectly observable" in some way that can be repeated.[36] Theories invoking "imaginary entities" such as atoms or genes are acceptable because atoms and genes are "indirectly observable" right now through experiments designed to test the consequences of atomic or genetic theory. In contrast, historical theories involve "*imagined events in the past* where no human beings were involved. . . . Such events as complete erosion of mountain ranges to sea level presumably resulting in peneplanes of extensive erosion surfaces that subsequently became buried due to further sedimentation, are totally imagined and unknown by any practicing, empirical scientists. These historical theories do not involve *natural* events."[37]

Moore asserts that, by definition, if events of the past cannot be observed or reproduced today, they cannot be natural. When theories about the past weave together genuine "natural" observations with "imagined events in the past," they become subscientific speculative conjectures because they rely on claims that are not fully "natural." In this fashion, Moore has redrawn the boundaries of science so as to exclude from the complex network of inference, theory, and observation the material that provides the foundation of the historical sciences.

Moore's conclusion that "historical theories do not involve *natural* events" is extraordinarily strong. It is the kind of claim that must lie behind the often repeated statement of Henry Morris that all "the real facts" of science support recent creation. But what is the warrant for Moore's conclusion? A nonstandard definition of what is "natural" is not adequate by itself. There may be significant distinctions between historical theories and theories about entities such as atoms and genes that can be probed by experiment—certainly our inability to test historical theories experimentally reduces the precision with which they can be stated—but the inability to experiment does not mean that historical theories are without substantial empirical support. If Moore is willing to acknowledge the legitimacy of theoretical claims about atoms and genes because they are "indirectly observable," he should likewise recognize that an analogous pattern of indirect confirmation also exists and is used

35. Moore, *How to Teach Origins,* p. 84.
36. Moore, *How to Teach Origins,* p. 89.
37. Moore, *How to Teach Origins,* p. 88; italics his.

to provide empirical evidence for the existence of the theoretical entities invoked by historical theories. One of the central pillars of contemporary astrophysics is the historical theory of stellar evolution. As Moore would quickly point out, neither the complete evolution of stars nor the nuclear fusion driving the process of stellar evolution is directly observable. But, as the case study in chapter 5 illustrates, many observable consequences can be inferred from stellar evolution theory, and when these inferences are tested (for example by plotting H-R diagrams for star clusters), the inferences are confirmed.

Actually Moore does seem to admit that historical theories involve indirect inferences from empirically observed events or objects:

> In attempts to explain *unobservable* origins of aspects of the present environment, scientists have formulated a very functional set of broad conceptualizations. Each of these broad conceptualizations involve identifiable observations or objects and/or events in the present life experiences of human beings, along with predictions primarily *after the fact* that are testable only by logical reasonableness or internal consistency regarding past events.[38]

His intent in objecting to "after the fact" predictions is not completely clear. It seems to assume that a theory such as stellar evolution, which is fundamentally a historical scenario constructed to account for specific observed properties of the physical world, cannot yield any predictions of observable events or processes except those that were originally utilized to construct the theory. In such a case the only grounds for accepting the theory would be logical reasonableness and internal consistency, and we justifiably would be rather leery of giving much weight to such a theory. But this is far from an accurate characterization of stellar evolution and other historical theories to which Moore objects. Stellar evolution has proved remarkably fertile as a theory, and it is that fertility combined with substantial cognitive relevance, external consistency, and unifying power that makes it a strong theory. Either Moore misses the point or is simply mistaken in his claims. In any case, they do not provide the kind of philosophical support creation science needs and Moore thought he had supplied.[39]

38. Moore, *How to Teach Origins*, p. 87; italics his.

39. Moore's analysis seems to be one of a family of similar attempts by creationists to deal with the problem of historical theories. The September 1984 issue of the *Journal of the American Scientific Affiliation* carried an article by Norman Geisler urging that we solve the problem by distinguishing between *operation* science, which deals with recurring patterns of events, and *origin* science, a much weaker form of analysis, which deals with

Not all creationists take such extreme positions. In *A Case for Creation,* Wayne Frair and Percival Davis provide a much briefer and more moderate discussion of historical inference. Under the heading "Prediction and Retrodiction," they simply note that because past events have had consequences in time, "It should be possible to erect testable hypotheses about events that may have occurred in the past. Then one could deduce the consequences and search for evidence that they had occurred."[40] They prefer to call such a process "retrodiction" rather than "prediction." As they point out, failed retrodictions can be explained away on the grounds that the ravages of history have destroyed the evidence that would have confirmed the retrodiction. Although the logical structures of prediction and retrodiction are identical in their perspective, retrodictions are weaker because their potential evidential basis is necessarily weaker. Thus they have a legitimate basis for concluding that "the scientific method is a practical device for resolving questions in 'real time.' It is far less effective when applied to the past."[41]

Retrodictions often may be stronger inferences than Frair and Davis are willing to concede, but it would be difficult to fault their broad point that in general they are weaker than "real time" predictions. The contrast to Moore and other creationists who employ arguments similar to Moore's could hardly be greater. And the contrast extends much beyond matters of content to include tone and style of argument as well. As the next two sections of this chapter suggest, these differences in tone

---

singularities. Henry M. Morris seems to prefer a more direct approach to solve the problem. In a short article, published as no. 10 of the ICR *Impact* series, he wrote,

> It should be quite obvious that . . . evolutionary processes cannot actually be observed. No astronomer has ever observed a "red giant" evolving into a "white dwarf," or a "spiral nebula" into a "globular cluster," or any other such change. This being the case, there is nothing whatever to prevent us from proposing the theory that they *don't* take place! This is by far the most reasonable theory, since it is supported by all the actual astronomic measurements that have ever been collected since man first began making such observations. If we limit ourselves to real, observational *science,* rather than indulging in philosophical speculation, we would have to say that the stars and galaxies have always been just as they are now since the time they were created. (P. ii).

There is something perversely impressive in such sweeping disregard of the entire theoretical structure of astrophysics and cosmology.

40. Frair and Davis, *A Case for Creation* (Chicago: Moody Press, 1983), pp. 17-18. Frair is a professor of biology at The King's College and a member of the Board of Directors of the Creation Research Society. Davis is a professor of life science at Hillsborough Community College, Tampa, Florida.

41. Frair and Davis, *A Case for Creation,* p. 21.

and style provide important clues to understanding important distinctions within the creation science movement.

TWO KINDS OF CREATIONISTS?

Perhaps there are two kinds of creationists. In contrast to the extreme claims made by creationists such as Moore, Morris, and Barnes, the comments by Frair and Davis are moderate and balanced. More importantly, Frair and Davis recognize and acknowledge both the strengths and the weaknesses of evolution and of creation science. In contrast to the insistence of Moore and many other creationists that the creation model is as good as the evolutionary model, and to Morris's claim that "all the real facts of science" support the creationist position, Frair and Davis write,

> We must not confuse the unaccustomed success that some have achieved in debate for real intellectual victory. We speak of the creation "model," but should not a model show a close correspondence to the reality it is supposed to represent? The construction of a detailed model, comparable in scope to the theory of evolution that we hope to replace, has hardly begun. Yet we cannot expect to compete effectively with evolution unless we construct such a model. The overwhelming detail with which evolution is replete contrasts discouragingly with the almost complete lack of positive discourse in creationism.[42]

The Frair-Davis assessment of the task yet to be accomplished gives every appearance of careful and realistic judgment, joined with a determination to lay the groundwork for substantial scientific achievement.[43]

A second look at the "shrinking Sun" articles published during 1980 and 1981 in the *CRSQ* reveals a similar phenomenon. There is a strong contrast in tone, style, and quality of argument between Steidl's article and those by Hinderliter and Hanson. Steidl provides a careful assessment of several lines of evidence that can be interpreted as raising

42. Frair and Davis, *A Case for Creation,* p. 139.

43. Frair apparently works at his science in addition to having written the general text *A Case for Creation.* Page 48 of that work displays a figure illustrating the use of zone electrophoresis of blood serum in taxonomy, and a note indicates that the figure is adapted from W. Frair, "Serological Survey of Pleurodiran Turtles," *Comp. Biochem. Physiol.* 65B (1980): 505-11.

questions about the existence of the solar fusion reaction. When new material appeared in the professional scientific literature that made his earlier argument less plausible, he immediately wrote a letter to the editor to warn readers of the *CRSQ* concerning the problematic aspects of his earlier claims. A significantly different rhetoric and strategy is evident in Hinderliter's contributions. Consider these excerpts, for example:

> Before ushering in the punch line, I should take the reader on a side excursion to deep, abandoned mines—for the sake of anyone who has read the literature, and wondered why in the world neutrino detectors should be located in such places.
>
> What does all of this have to do with the shrinkage of the sun anyway? To be blunt, solar contraction is the refreshing breeze that simply blows away the foggy dilemma.
>
> I thought to myself, "Hogwash! Is that the only reason why they discounted gravitational contraction?"[44]

The Hanson article published a year later has the style of a scientific article, but its assumptions and methodology are extremely simplistic.[45] The remainder of the creationist literature dealing with the shrinking Sun merely presents as well-established results what the larger scientific community has abandoned as a mistaken enterprise.

How can we account for the striking contrast between the judicious assessment of the status of creationist achievements given by Frair and Davis and the extravagant claims of Morris, Barnes, and Moore? Why is Steidl able to recognize and acknowledge the limitations of his initial discussion of the Eddy and Boornazian paper while Hinderliter and Hanson seem to be unable to deal critically with their treatment of the same material? The differences seem to involve more than a matter of personal style. In the remainder of this chapter we will explore the possibility that while Steidl, Frair, and Davis are attempting to develop a legitimate creation science, the work of Morris, Barnes, Moore, and others like them fails to meet even minimal standards for responsible scientific research and publication—that their very considerable energies are engaged in a related but quite different kind of effort.

44. Hinderliter, "The Shrinking Sun: A Creationist's Prediction, Its Verification, and the Resulting Implications for Theories of Origins," *CRSQ* 17 (1980): 58, 59.

45. For a discussion of the shortcomings of Hanson's paper, see *Science Held Hostage,* pp. 60-62.

WORLDVIEW, SCIENCE, AND FOLK SCIENCE: DEVELOPING A THEORETICAL FRAMEWORK FOR UNDERSTANDING CREATIONISM

What is the relationship between science and religion? Recent studies in the history of science have led many historians to reject the warfare metaphor that dominated many earlier discussions of science and religion. Indeed, a recent popular account of these studies emphasizes the mutually beneficial "cross-currents" that have repeatedly linked the fortunes of science and Christianity.[46] On the other hand, the line we have drawn between the domain of science and the domain of religious inquiry is so sharp that it is difficult to understand how any substantial and legitimate interaction is possible. Perhaps this is because the whole point of the "domain" concept is to develop a set of categories that will help us identify and deal with abusive behavior when inappropriate claims are made in the name of science or religion. Its fruitfulness in that respect is suggested by the helpful discussion of the formation-origin and behavior-governance distinctions. Now we need to pay more attention to the ways in which the distinct—almost nonintersecting—realms of science and religion might interact with one another.

In chapter 5 we briefly note two legitimate modes of interaction between science and religion. One of these occurs at the highest cognitive level: in the struggle to build and maintain a comprehensive worldview, we suggested that is important to learn from both science and religion "so that we may come to know not only the intrinsic intelligibility of the physical universe but also its place within the whole of reality." On a lower cognitive level, science and religion interact if a scientist decides to carry out a particular program of research under the stimulus of religious commitments. We suggested that this is permissible and legitimate as long as appropriate standards of craft competence, professional integrity, and epistemic judgment are maintained. But if science and religion deal with such different domains, how is it possible to bring them together as we attempt to build a comprehensive worldview? And is it creditable to suggest that a scientist should be able to respond to the stimulus of religious commitment in the design of a research program and still manage to maintain appropriate standards of craft competence, professional integrity, and epistemic judgment when evaluating the results of that research program? An exploration of these questions will

46. Colin A. Russell, *Cross-Currents: Interactions between Science and Faith* (Grand Rapids: Eerdmans, 1985).

help provide a broader foundation for understanding the creationist movement.

We would do well to begin our exploration with a consideration of James Olthuis's view on the role of worldviews as mediators between life and faith. He writes that worldviews serve simultaneously as "visions of faith for life [and] visions of life for faith."

> As a vision rooted in faith . . . a worldview first shapes itself to faith and then shapes the world to itself, projecting images of cosmic order on the plane of human experience. However [that] is only half of the story, the half told from the stance of faith. From the other direction—from the rest of life experience—comes the other half of the story. . . . In the movement from life experience to faith experience, a worldview first shapes itself to the world and then shapes faith to itself, atuning and adjusting images of the cosmic order so that they mirror experienced reality.[47]

As a mediator between faith and life, a worldview provides a vantage point—the perspective—that actively shapes our understanding of the world *and* our faith. Yet a worldview, *if it is to remain viable,* must be open to revision "as faith deepens [and] as insight into reality grows."[48] This is a difficult challenge, especially in a world that is changing as rapidly as ours. Because they are human creations, our worldviews can never be fully adequate, and yet we are bound to accept as true the perception of reality they provide for us. Olthuis's insights have the potential to shed a great deal of light on the creation science movement, but they need to be augmented by several concepts drawn from the work of philosophers and sociologists of science.

First, let us consider Jerome Ravetz's concept of "folk science" as a means of understanding the process through which a worldview mediates between world and faith. Ravetz argues that science becomes folk science when it functions in a worldview to provide its adherents "comfort and reassurance in the face of the crucial uncertainties of the world of experience."[49] And he contends that the professional scientist is just as likely as a layperson to adopt a folk science.

Einstein provides us with a clear example. He was quite aware of the extent to which he found security and comfort in his communion with the extrapersonal world of physics where "wishes, hopes and primitive feelings" did not reign. He explicitly saw this as an alternative

47. Olthuis, "On Worldviews," *Christian Scholar's Review* 14 (1985): 158-59.

48. Olthuis, "On Worldviews," p. 163.

49. Jerome R. Ravetz, *Scientific Knowledge and Its Social Problems* (New York: Oxford University Press, 1971), p. 386.

to the religious solace he had sought as a youth.[50] And Einstein is far from unique in this respect. A more popular form of folk science recently found a prominent place on our TV screens as Carl Sagan intoned "The Cosmos is all that is or was or ever will be."[51] Einstein and Sagan are joined by countless other adherents of a wide variety of folk sciences (e.g., deism, evolutionism, Paleyan natural theology, eighteenth- and nineteenth-century faith in the long-term benevolence of scientific and technical progress), and the resonance of all of them is clearly religious.

There is nothing inherently disreputable about folk science, but folk sciences bear watching because of the intellectual and religious mischief they may produce. Folk science provides a standing invitation to the unwary to confuse science with religion (à la Sagan) or to allow the religious perspectives present in the folk science to feed back into the scientific world to distort its development. It is just this latter process that creation scientists say has allowed evolutionism to derail much of modern science, while many who object to creation science repay the compliment in their dismissal of creationist claims as thinly veiled religious advocacy.

There is an interesting symmetry in the claims and counterclaims of evolutionists and creationists. Both sides have legitimate cause for complaint. The key to understanding their complaints lies in a clarification of the different criteria of usefulness and adequacy that link professional science and folk science to the knowledge base they both share but use in very different ways for quite different purposes.

Professional science uses its knowledge base to improve our understanding of the natural world. In that process the products offered by scientists are assessed in terms of their *usefulness* for the further development of the scientific field to which they belong and their *adequacy* with respect to the professional standards and the relevant craft knowledge of that field. In contrast, the products of folk science are accepted quite independently of their achievements in solving scientific problems or producing secure knowledge of the world. These products are judged *useful* to the extent that they address social, emotional, or religious problems central to the concerns of the audience, and their *adequacy* is assessed by their "success in offering reassurance and comfort."[52]

50. Ravetz, *Scientific Knowledge and Its Social Problems,* p. 66.

51. Sagan repeats the statement in the book published as a companion to the series, *Cosmos* (New York: Random House, 1980), p. 4.

52. See Ravetz, *Scientific Knowledge and Its Social Problems,* p. 388.

How does all of this relate to Olthuis's views? In keeping with his view of the ways in which a worldview acts as a mediator in the two-way movement between faith and world, we might say that in the movement from faith to world, the criteria of usefulness and adequacy by which the view of the world is assessed are those of folk science, whereas in the movement from world to faith, the criteria of usefulness and adequacy by which the view of the world is assessed are those of professional science. For this reason folk science and science need to find a means of "honest" mutual accommodation within a worldview. *Such an accommodation is achieved only when there is an internal dialogue as the vision of folk science is critiqued by science and the assumptions of science are critiqued in turn by folk science.*

In an honest dialogue, each side must be open to learning from the other. When the result of such a dialogue is a sense of mutual support, great cultural energy can be released—as it was, for example, in early encounters with deism, with Paleyan natural theology in the first third of the nineteenth century, and with the idea of scientific and technical progress throughout the nineteenth century, and as it is with evolutionary naturalism today. If the outcome is a sense of conflict, the result is uncertainty and confusion—as in the case of the decline of evangelical scholarship at the turn of the century.[53] When a pseudoagreement is imposed by allowing folk science to dominate science, science loses its capacity to extend and deepen our understanding of the intrinsic intelligibility of the created world. If science is allowed to dominate folk science, the result is a scientism that oversteps the domain about which science is competent to speak. In effect, scientism creates a folk science while denying the existence of the distinction between science and folk science. In either case the possibility of internal dialogue withers and the worldview loses its capacity for revision and virtually demands its canonization as an "infallible blueprint for life."[54] This gives rise to very serious problems. Olthuis contends that

> Such absolutization of our views conveniently absolves us from the need constantly to test and refine our own perceptions, and it negates the possibility of seriously considering any other perception of truth and reality. Indeed, it blocks us from being truly open to God's revelation. . . . If we canonize our worldviews or even if we adopt static worldviews,

53. On this, see George Marsden, "The Collapse of American Evangelical Academia," in *Faith and Rationality: Reason and Belief in God,* ed. Alvin Plantinga and Nicholas Wolterstorff (Notre Dame, IN: University of Notre Dame Press, 1983), pp. 219-64.

54. Olthuis, "On Worldviews," p. 162.

> the development of faith and the development of insight in the light of faith are stopped cold.[55]

As we try to shape a worldview responsive to both science and religion, it is important to recognize the role played by folk science in mediating between faith and science even as we recognize its essentially religious character.

Our second step in developing an understanding of the process through which worldview mediates between world and faith involves a closer examination of the concept of *adequacy*. In particular we need to focus on the broad classes of criteria used by scientists in assessing the adequacy of their work and the work of others. Within science the concept of adequacy is rooted in the following claims:

1. Theories of science are less than fully determined by the empirical evidence that supports them.
2. Because theories of science are less than fully determined by the empirical evidence that supports them, individual scientists can achieve no more than adequate solutions to their research problems.
3. Criteria of adequacy are not given by nature but grow out of the collective experience of the research community to which scientists belong.
4. Most criteria of adequacy used by scientists exist only as tacit knowledge.[56]

Craft knowledge and professional practice are two aspects of scientific work in which criteria of adequacy play a crucial role; in both instances they serve to define what counts as acceptable or competent practice. In chapter 5 we emphasize the importance of craft competence as part of the value system of science. We are now extending that discussion by pointing out the role of criteria of adequacy in determining what counts as competent practice. Each discipline and subdiscipline has a set of criteria (some are explicit but most are tacit) that bracket acceptable practice. These criteria range from dicta concerning appropriate research

55. Olthuis, "On Worldviews," p. 162.

56. For more information on the concept of adequacy, see Ernan McMullin, "Values in Science," *PSA 1982: Proceedings of the 1982 Biennial Meeting of the Philosophy of Science Association,* vol. 2, ed. Peter D. Asquith and Thomas Nickles (East Lansing, MI: Philosophy of Science Association, 1982), p. 13. For a discussion of tacit knowledge, see our chap. 5, pp. 146-47.

strategies, instrumentation, and materials to acceptable patterns of data assessment, inference, and argument. These criteria are learned initially by graduate students through immersion in research under the tutelage of professional scientists. Such knowledge is sustained and broadened throughout the course of a scientific career only through continuous participation in the work of a research community.

The practice of professional integrity is analogous to craft competence. We have argued that science cannot flourish unless scientists maintain high professional standards of honesty, fairness, and candor. They learn what counts as honest reporting and fair treatment of conflicting evidence and the like in the same way they learn about craft competence and adequacy—through immersion in research under the tutelage of experienced researchers and subsequent participation in a research community. Scientists are expected to take reasonable precautions to ensure that their results are reliable, and a determination of what counts as "reasonable" is worked out in the give-and-take of a continuing research program within a particular field. Maintaining a high level of professional integrity within a field of research is largely the responsibility of its senior members, who provide examples of acceptable behavior in their own work and in their service as editors, reviewers, and mentors of new entrants to the field. An area of research cannot flourish without a high level of adherence to appropriate standards of scientific integrity.

CREATIONISTS REASSESSED

While the line of division may not be uniformly sharp and crisp, there are two kinds of creationists. The first—and smaller—category includes individuals such as Frair, Davis, and Steidl, whose writing gives evidence that they have been able to establish and maintain an internal dialogue allowing science and folk science to critique one another. Their creationist worldview clearly plays a role in shaping their research agenda, in heightening their interest in particular research areas, and in increasing their awareness of the problematic aspects of contemporary science. But they have retained the ability to recognize when the connections they have forged between science and folk science have failed and to adjust their stance accordingly. We might call them "modest scientific creationists."

The second category might be characterized as "immodest

populist creationists."[57] It includes most of the leaders of the creation science movement. Barnes, Morris, Moore, and Hinderliter are certainly members. In their work, folk science has taken control, and the internal dialogue between science and folk science has essentially ended. Criticism from the world of professional science is explained away as the result of evolutionary indoctrination or mere speculation. Hinderliter provides a good example of this in his assessment of the evidential support for the gravitational collapse theory of solar energy: "By this time I had done considerable study on the scientific evidence bearing on the age of the earth, solar system, etc.; and I had come to realize that the compelling force for the acceptance of vast ages was merely a faith in evolutionism, which itself has no evidential leg to stand on."[58]

Creation scientists in this category write mainly for one another and the widespread grass-roots lay audience that supports the creationist movement. According to Henry Morris, the most "visible" publication of the ICR is its monthly newsletter, *Acts and Facts,* which is distributed free of charge to their 75,000-address "active" mailing list. Each issue of *Acts and Facts* is accompanied by a 1,500- to 2,000-word *Impact* article, copies of which are also widely distributed through seminars, churches, and public debates. The *Impact* articles are by design popular essays by creationist leaders. They are lively, well written, and highly quotable because they are filled with strong and unqualified

57. Whether or not it is helpful to consider creationism a populist movement should be the subject of further research. Creationism has a number of characteristics that suggest this would be a helpful line of inquiry. Most importantly, creationism seems to have arisen in many different places virtually simultaneously. In keeping with the classic pattern of populism, it is a reaction to something that is deeply threatening to the belief system (or way of life) of its adherents and over which they have very little control. In addition,

- its leaders remain, by choice or by necessity, on the margins of the professional world they claim to represent;
- its central institutions were organized explicitly as counterinstitutions;
- its major strategy for engaging the world of scholarship is public debate and popular lecture rather than publication in scholarly journals;
- it attempts to circumvent the scientific establishment by appealing to the general public, to school boards and state legislatures, and by using the legal system to establish its "science" in the public classroom; and
- it devotes great energy to fostering grass-roots support.

58. Hilton Hinderliter, "The Shrinking Sun," p. 59. [EDITOR'S NOTE: In a letter to the editor of *Origins Research* (12 [1989]: 3), Hinderliter appears to have had second thoughts about this assessment and offers an example of how he would now prefer to speak more objectively: "If I were writing today about surprising new evidence that indicated a rapid shrinking of the sun, I would conclude, '*If* the observations stand up to continued analysis, it *could* mean that the sun derives its energy from gravitational contraction.' I say this not because the final word is in on solar contraction, but because it was the most correct and logical conclusion to have been drawn all along."]

claims. They are meant to have an impact on their readers, and there is every reason to believe that they do. An *Impact* article by Thomas G. Barnes provides a clear example of the genre. Under the heading "Oceans of Piffle in Evolutionary Indoctrination," he writes,

> The application of evolutionary doctrine to the origin of the universe always involves inverted logic. It is typified by the so-called big bang theory of the origin of the universe. According to that dominant theory, the universe began as a ball of energy and evolved, through the processes of explosion and expansion, into our highly ordered and beautiful universe. If there ever was an inversion of logic, that is it. An explosion does not produce order, it produces disorder! The big bang theory violates all the applicable governing principles of physics. The multitude of papers espousing the big bang explanation of origins are indeed nothing more than piffle. In case the reader is unfamiliar with the word piffle, Webster's Dictionary describes it as "trifling talk"; "stuff and nonsense"; "twaddle," all of which applies to the inverted logic in evolutionary indoctrination.[59]

Clearly these assertions fall into the category of folk science. Is that also true of the articles Barnes, Morris, and Moore publish in the *CRSQ*? Typically their format and style are different from those of the short *Impact* articles, but are the controlling criteria of value and adequacy appropriately different? This crucial question can be settled only by a thorough analysis of creationist literature, but at the very least we have uncovered serious problems in this area.

For one last illustration of the problematic side of creationist research, we turn again to a chapter in Morris's *History of Modern Creationism* in which he describes the development of the Institute for Creation Research. Established in 1972, it was the first creationist institution to have a full-time staff "dedicated to research, writing, and teaching in the field of scientific creationism." Funding for creationist research has been difficult to secure, and Morris notes that the "lack of large research programs . . . has been a weakness of the creation movement." On the other hand, he claims that "some worthwhile creation-oriented research" has been accomplished. But what research is he referring to, and what criteria did he use in characterizing it as "worthwhile"? He emphasizes that ICR research has been concentrated "wherever the potential impact in terms of Christian witness would be greatest in relation to expenditures." This has led to a focus on three

59. Barnes, "Oceans of Piffle in Evolutionary Indoctrination," *Impact* no. 142, Institute for Creation Research, April 1985.

areas—overthrust faults in "Colorado, Nevada, and other places"; anomalous fossils in "Utah, Oklahoma, and other places"; and explorations of Mount Ararat. Research in these area has not led to publications, he explains, because while "the evidence seemed strongly to favor the creation/flood model, [it] was not sufficiently compelling by itself to settle the question." The two Ararat expeditions "have found a number of interesting archeological sites, but still no signs of Noah's ark."[60]

The only claimed contribution to creation science stemming from the major thrusts of the ICR research program has come from studies of various sites along the Paluxy River in Texas, where "human footprints have been reported to be associated with dinosaur footprints." These studies are reported in *Tracking Those Incredible Dinosaurs and the People Who Knew Them,* which is described as "an excellent research monograph." In fact, the publication is a reasonably careful popular account, but it scarcely qualifies as a research monograph.[61]

60. Morris, *History of Modern Creationism,* pp. 250-53.

61. *Tracking Those Incredible Dinosaurs and the People Who Knew Them* (San Diego: Creation-Life Publishers, 1980) was written by John D. Morris, the son of Henry Morris. It consists of approximately 30,000 words of text and about seventy pages of charts and photographs. There also is an appendix entitled "Detailed Descriptions of the Major Human Trails," which consists of approximately 6,000 words of text and about twenty-three pages of pictures and tables. The entire publication has sixty-five footnotes, of which thirteen are citations of the professional geological literature, three refer to analytical results from commercial laboratory reports, one cites a *Creation Research Society Quarterly* article, one cites a standard engineering text, and seven cite books by creationist authors. The remaining citations refer to unpublished manuscripts (fifteen), popular science books or articles (ten), private correspondence (two), and popular creationist literature (thirteen). A chapter entitled "Proper Identification of the Tracks" contains only one reference to the professional scientific literature (a geological field guide). The pattern of citation is appropriate for a careful popular work, but hardly what is necessary for a "research monograph."

During the 1980s a number of creationist and anticreationist researchers examined the sites that provided the data for the Morris monograph. In the January 1986 issue of *Impact* (no. 151, "The Paluxy River Mystery") John Morris describes some of the new findings that called into question his earlier claims. Although calling for continued research, he warns fellow creationists that "it would now be improper for creationists to continue to use the Paluxy data as evidence against evolution" (p. iv). To my knowledge this is the first time that an important creationist claim has been even partially retracted in material published by the Institute for Creation Research. A critique of the limited character of the Morris retraction can be found in Ronnie Jack Hastings, "New Observations on Paluxy Tracks Confirm Their Dinosaurian Origin," *Journal of Geological Education* 35 (January 1987): 4-15. Hastings is one of the most active of the anticreationists investigating the Paluxy River sites. In the article just mentioned and in three others ("Tracking Those Incredible Creationists," *Creation/Evolution* 15 [1985]: 16-36; "Tracking Those Incredible Creationists—The Trail Continues," *Creation/Evolution* 17 [1986]: 19-27; and "Tracking Those Incredible Creationists—The Trail Goes On," *Creation/Evolution* 21 [1987]: 30-42), he has provided a highly colored but detailed account of research findings since 1980 and the interactions of creationist and anticreationist researchers interested in the Paluxy River sites.

These are rather modest "scientific" accomplishments for more than a decade of effort. In fact, by any normal measure of scientific work, unpublished field notes, travel adventure stories, and popular accounts are not scientific accomplishments at all. On the face of it, what seems most strange is their stated decision not to publish the results of their field studies because the evidence was not sufficiently compelling to settle the question decisively in favor of the creation / flood model. If the goal of this work was to further creation science (a professional goal), then even less than overwhelmingly favorable results would merit publication. After all, who would expect the evidence collected on a number of field trips to decisively overturn what the broader geological community has long accepted as a firmly established complex of observation, law, theory, inference, and model? The decision not to publish would make more sense if the goal of the field trips was mainly to produce results that would reinforce the folk science aspects of creationism. In fact, Morris virtually stipulates that folk science was the reigning goal in his description of the criteria governing the choice of research projects. The choice was made to concentrate "research efforts wherever the potential impact in terms of Christian witness would be greatest."[62] One can sympathize with the goal, but at the same time it is important to recognize the way in which folk-science values are shaping the agenda, research strategy, and publishing decisions of an ostensibly scientific enterprise.

At its core the contemporary creation science movement is an attempt to provide a scientific foundation for a folk science that for years had been floating free of any support within the professional scientific community.[63] Because the leading creation researchers seem committed to preserving with only cosmetic changes the folk-science superstruc-

62. Morris, *History of Modern Creationism*, p. 251.

63. For an excellent short history of the creationist movement, see Numbers, "The Creationists." The crucial catalyzing event of the contemporary phase of the movement was the publication of Henry M. Morris and John C. Whitcomb's *The Genesis Flood* (Philadelphia: Presbyterian & Reformed, 1961). In it a Ph.D. engineer (Morris) collaborated with a theologian (Whitcomb) to produce what was essentially an updated version of George McCready Price's flood geology. In his *History of Modern Creationism* Morris says the agreement to cooperate came after he reviewed a manuscript of Whitcomb's in which the scientific chapters "were essentially merely a survey of George McCready Price's arguments. This was all right as far as it went, but Price himself had failed to make much of an impact with these same arguments 30 years previously so I suggested that a new approach was needed, with better and more recent documentation" (p. 150). But the crucial difference lay not in the content but in the fact that for the first time in eighty years or more a person trained at the doctoral level in what appeared to be a relevant technical discipline (hydraulic engineering) had collaborated in the publication of a full-scale argument advocating a young Earth and a worldwide flood.

ture, they are faced with the prospect of having to alter the scientific foundation to fit what is already in place. Perhaps this is the greatest barrier to increasing the quality of the science within the creation science movement.

CONCLUDING COMMENTS: UNDERSTANDING THE CREATION SCIENCE MOVEMENT

Implicit in the material we have presented in this chapter are three fundamental questions that lie at the heart of any assessment of the creation science movement:

1. In the abstract, is the idea of a research program devoted to the development of creation science conceivable?
2. Are there any creation scientists?
3. Does a sufficiently coherent body of results exist to justify the claim that there is a creation science that can be taught in the schools?

To review briefly, the answers we have given in the preceding pages are:

1. Yes, research devoted to the development of creation science is conceivable if we recognize that the kind of science sought by members of the contemporary creationist movement is a "sectarian science" because creationists insist that the controlling epistemic value governing science is that all explanations must be in accord with their understanding of the early chapters of Genesis.
2. Yes, there do seem to be a few genuine creation scientists. We would point to Frair, Davis, and Steidl, who apparently have been able to establish and maintain an internal dialogue within which science and folk science critique one another.
3. No, the results of creation science research are neither voluminous nor coherent. Even creationists such as Frair and Davis have noted the "almost complete lack of positive discourse in creationism," and Henry Morris himself is unable to claim anything beyond the most modest successes for twelve years of research supported by the Institute for Creation Research. If we add to these self-acknowledged limitations the very serious prob-

lems associated with many creationist claims of proof that the Earth is young, then it seems completely unreasonable to claim that anything resembling a legitimate creation science exists.

We freely acknowledge that these answers are controversial because on the one hand there isn't any general agreement among scientists, philosophers, historians, or sociologists of science about the essential attributes of scientific activity, and on the other hand the creation science movement is complex enough that no one person is fully competent to assess it. Nevertheless, these questions need to be addressed, and our intent in this chapter has been to suggest a way of bringing order to what has often been a rather disorderly and confused arena of conflicting claims and counterclaims.

What should creation scientists do? Because creation science uses Scripture to establish its controlling epistemic value (the conclusions of creation science must agree with a literal reading of Genesis 1–11), it has developed as an isolated sectarian science. And because the creation science program implicitly demands the restructuring of much of modern science, it places extraordinary demands on its adherents. Cut off from the normal sources of funding by their epistemic deviance, and institutionally isolated from the normal channels of publication and criticism that are essential to the healthy development of a scientific field, creation science researchers are faced with the staggering task of developing a radical position in the face of substantial opposition while at the same time subjecting that position to a searching critique. It is not easy to do creation science with integrity.

The sectarian isolation of creation science is amplified, and the possibility of maintaining high internal critical standards is undermined, for those creationists who adopt anything like Moore's philosophy of science or Henry Morris's biblical hermeneutics. The Morris hermeneutic combines a resolute blindness to the context within which Scripture was given with the assumption that any biblical reference to the created order should be taken as a straightforward factual statement relevant to contemporary science.[64] Add Moore's philosophy of science to Morris's

64. Products of this hermeneutic permeate Morris's writings. An analysis of his interpretive strategies would make an interesting study. Following are examples drawn from his *Studies in the Bible and Science* (Philadelphia: Presbyterian & Reformed, 1966):

> (1) "If one starts with the presupposition that God has written the Bible as His own perfect revelation of the origin, purpose and destiny of the world, then it . . . is perfectly possible to correlate all the physical data of science and history within that framework" (p. 109).

hermeneutic, plus the creationist insistence that the Earth must look young, and the result is a virtual license for poorly controlled speculation: the combination simultaneously dismisses much of what is assumed within mainstream science to be established results that must be taken into account if a theoretical proposal is to be taken seriously, and it invites a kind of anything-goes (my speculation is as good as your speculation) approach to discussing the consequences of various formative history scenarios.

---

(2) "Our concern here is simply to show that the Bible does provide a perfectly sound basis for understanding not only religious truth but also physical processes. It may very effectively serve as a 'textbook' of scientific principles within which we can satisfactorily explain all the data of science and history" (p. 110).

(3) "It is quite impossible for man, with his study of *present* processes, to know anything for certain about the prehistoric past. . . . Only God can *know* these things, and we are able to know the truth about these matters only through faith in God's statements concerning them" (p. 110).

(4) After quoting Hebrews 1:2-3 Morris says, "Thus, by Power, by the Word, all things were made, and all things are upheld. Jesus Christ, through the continual outflow of His limitless divine energy is thus sustaining all of the material stuff of the universe which he had once created. Here is clearly spelled forth the modern scientific truth of the equivalence of matter and energy. Here also is revealed the ultimate source of the mysterious nuclear forces, the binding energy of the atom" (p. 113).

(5) "If men had been willing to develop their scientific systems on the basis of Biblical presuppositions . . . it should have been quite obvious all along that the basic physical processes were those of conservation and decay, as now formalized in the statements of the first and second laws of thermodynamics" (p. 114).

(6) Speaking of Genesis 2:1-3, Morris states that "This statement is as clear as it could possibly be in teaching that God's creative acts were terminated at the end of the six days. . . . It is therefore quite impossible to determine anything about Creation through a study of present processes, because present processes are not creative in character. If man wishes to know anything at all about Creation . . . his sole source of true information is that of divine revelation. . . . Therefore we are completely limited to what God has seen fit to tell us . . . in His written Word. This is our textbook on the science of Creation" (p. 114).

(7) Morris contends that the second law of thermodynamics describes "a universal tendency toward decay and death" that strikes us as being "undesirable and abnormal in a universe created by a Holy and Omnipotent Creator." How are we to understand this thermodynamic principle? He maintains that it "is all explained and long anticipated in Scripture, which attributed it to the entrance of sin into the world. . . . As the Scripture says: 'the whole creation groaneth and travaileth in pain together until now' (Romans 8:22). The whole world, both the heavens and the earth, and all that is in them is, are 'waxing old, as a garment' (Hebrews 1:11)" (p. 115).

(8) Speaking to the concerns of meteorology, Morris says, "The essential character of the atmospheric heavens as a sort of terrestrial blanket which makes the earth inhabitable, retaining the heat and spreading the light from the sun, as well as

What should we ask of people who claim to be creation scientists? It is unreasonable to ask them to modify their ruling epistemic value—in that case they would no longer be creation scientists. But it does seem appropriate to suggest that if other Christians, let alone other scientists, are to take their claims seriously, they need to continue to pay their professional dues through research and publication in the usual professional journals. If there is any substance to the creationist vision of the world, research programs based upon it should generate a stream of observations that are strikingly anomalous with respect to expectations generated by the standard assumptions about the formation and age of the Earth. These observations should be readily publishable in standard journals because they would raise significant questions—as the work of Eddy and Boornazian which set off the shrinking Sun episode raised significant questions. But with one notable exception, creationist authors have not been able to uncover and publish the strikingly anomalous phenomena that their theories would lead one to expect. The exception is the work of Robert Gentry, who has identified what appear to be halos in granite crystals caused by the radioactive decay of polonium 218. The significant questions raised by one of Gentry's papers was summarized in a *Geotimes* report:

> The polonium . . . halos are the center of a mystery. The half life of the isotope is only 3 minutes. Yet the halos have been found in granitic rocks at considerable depths below land surface, and in all parts of the

---

providing air to breathe, is pointed out in Isaiah 40:20 which says that God has 'stretched out the heavens as a curtain, spreading them out as a tent to dwell in' " (p. 118).

(9) On the biblical principles applicable to geology Morris states that "The basic principle of isostasy (meaning 'equal weights'), which is the foundation of geophysics, is indicated by Isaiah 40:12, which speaks of God 'weighing the mountains in scales, and the hills in a balance,' from which the pre-eminent importance of gravitational forces in geophysical calculations should easily be inferred" (pp. 118-19).

(10) "Assuming the Biblical revelation to be true, there is no reasonable conclusion possible other than that the Flood must serve as the main vehicle of interpretation in developing a valid science of historical geology" (p. 119).

(11) "The science of *physiology* can well be understood as centered around the wonderful Biblical statement that 'the life of the flesh is in the blood' (Leviticus 17:11)" (p. 120).

(12) "In similar fashion, one could examine the sciences of *archeology, anthropology, medicine, taxonomy,* and others, and again and again one could find that the basic principles are revealed in Scripture. . . . The Bible is, for the Christian, *the* textbook of science and of all knowledge" (p. 120).

> world. . . . The difficulty arises from the observation that there is no identifiable precursor to the polonium: it appears to be primordial polonium. If so, how did the surrounding rocks crystalize rapidly enough so that there were crystals available ready to be imprinted with radiohalos by alpha particles from ^{218}Po?[65]

Although some of Gentry's work has received extensive criticism in educational journals,[66] his puzzling claims have not yet been fully resolved in the professional literature. But, as Thomas Kuhn emphasized more than twenty-five years ago, the existence of anomalous results such as the Gentry observations is a normal state of affairs within virtually every branch of science.[67] The existence of a few such anomalies is not enough to convince scientists that there are serious problems with a theory or a field of research. Perhaps the most telling weakness of creation science has been its inability to discover and establish the existence of a multitude of anomalies that cannot be explained by the theories with which creationists disagree.

The failure of creation science stems from two related problems. As an immature science (i.e., a science without any examples of successfully solved problems), creation science does not support its practitioners with the kind of guidance provided by tested and effective

65. Raphael G. Kazmann, "It's about Time: 4.5 Billion Years," *Geotimes* 23 (Sept. 1978): 19; cited by T. G. Barnes, *CRSQ* 23 (March 1987): 168.

66. See Jeffery Richard Wakefield, "The Geology of Gentry's 'Tiny Mystery,'" *Journal of Geological Education* 36 (1988): 161-75. Wakefield provides extensive evidence that Gentry's samples were not taken from "basement rocks" or "primordial rocks," but from intrusive dikes that formed much later. In his closing paragraph Wakefield concludes that

> The geology of the sites at which Po halos are found clearly shows that Gentry's proof of instantaneous creation and a young Earth is nothing of the sort. Gentry's Po halos simply do not occur in "primordial granites," but instead were formed in relatively young dikes that demonstrably crosscut older sedimentary and igneous rocks. Gentry claims to be an objective scientist but he has, in fact, ignored the very extensive published evidence that disproves his hypothesis. In addition, when confronted with this evidence he simply denies its existence. Such behavior is not characteristic of scientists, but of pseudoscientists.

[EDITOR'S NOTE: For a recent article in the professional research literature, see A. Leroy Odom and William J. Rink, "Giant Radiation-Induced Color Halos in Quartz: Solution to a Riddle," *Science* 246 (1989): 107-9. Odom and Rink offer a solution to Gentry's "riddle" that "requires neither unknown radioactivity nor an abandonment of current concepts of geologic time."]

67. Kuhn, *The Structure of Scientific Revolutions*. This book has played a major role in setting the agenda for a very vigorous discussion of the nature of scientific knowledge and the process of scientific change—a discussion that is still vital and that has led to dramatic changes in our understanding of these areas. Few would question Kuhn's claim that anomalous results can be found in many successful fields of research.

research programs.[68] It is a demanding task to do good science even in a well-established field. It is extremely difficult within the context of an immature science. It is even more difficult when the immature field is called upon to provide the solution to a pressing social problem. In a very real sense that is the situation confronting creation scientists. They are attempting to practice in a field that exists to provide an answer to the challenge of evolutionary thought to Christian faith. But the proposed solution, creation science, has not been able to establish a set of stable answers.

Jerome Ravetz has suggested that, in a situation of this sort, two things are likely to happen. First, in seeking public support for their views, leaders of the field are likely to conceal its condition "from themselves as well as from their audience."[69] If they believed their field to be mature when in fact it is immature, then "the social mechanisms for quality control and direction in the field cannot function properly, the safeguards against the abuses of prestige are weakened, and the assessments of a lay audience, based on popularization, can be of more practical importance in the politics of the field, than those of the community of experts."[70] In many ways this seems to be an accurate description of the current situation in creation science.

The second thing likely to happen in such a situation, "where facts are few and political passions many," says Ravetz, is that "the relevant immature field functions to a great extent as a 'folk-science.' This is a body of accepted knowledge whose function is not to provide the basis for further advance, but to offer comfort and reassurance to some body of believers."[71] There can be little doubt that creation science functions as a folk science for lay readers of the ICR *Impact* series as well as for the readers of other creationist publications such as the *Bible Science Newsletter*.

Our discussion of the relationship between worldview and folk science leads to the conclusion that creation science also functions as a folk science within the worldview of those individuals attempting to establish creation science as a research program. For creation scientists such as Frair, Davis, and perhaps Steidl who are able to recognize the

68. To the extent that creation scientists rely on flood geology as an example of successful creation science, they are misled. See Davis Young's case study of creationist claims concerning the geology of the Grand Canyon published in *Science Held Hostage*, pp. 93-124.

69. Ravetz, *Scientific Knowledge and Its Social Problems*, p. 378.

70. Ravetz, *Scientific Knowledge and Its Social Problems*, p. 379.

71. Ravetz, *Scientific Knowledge and Its Social Problems*, p. 366.

limited results achieved thus far by creation science (and thereby implicitly recognize its status as an immature field), there is evidence that the internal dialogue of mutual criticism between folk science and professional science is alive and thus that their worldview is healthy. Their achievements in creation science may be modest, but they remain *scientists* attempting to develop a research program under difficult circumstances. But within the world of creation science, they are overshadowed by such figures as Morris, Barnes, and Moore, who are unable to recognize the immaturity of their research program and whose publications suggest that the essential internal dialogue between folk science and professional science no longer exists. Without realizing what they have done, many of the leaders of the creation science movement have betrayed the trust placed in them by their lay followers. The pervasive lack of critical judgment that characterizes the creation science literature is due to its role as a folk science intended primarily to offer "comfort and reassurance to believers" rather than to make a contribution to our deeper understanding of the created world. The dominant figures in the creation science arena function not as scientists but as populist religious leaders whose authority and prestige are rooted in the folk science they proclaim.

7. WHAT SAYS THE SCRIPTURE?

JOHN H. STEK

Unavoidably we bring to our reading of Scripture our present-day concerns, our modern Western vocabulary, and our contemporary concepts of literature. We also draw from a particular received tradition concerning what the Bible authoritatively teaches and what Scripture, thereby, requires us to believe and to do. Sometimes what we bring to Scripture is very helpful as we seek a faithful understanding, but in other instances it may cloud our perception of what is actually in the text and may lead us to see something quite alien to the original meaning. Hence we appreciate the efforts of biblical scholars and theologians who provide us with the means for becoming better informed readers of God's written Word.

In the context of our present study, several questions stand out as particularly important: What is being taught or revealed in Genesis 1? Are the details of this narrative (such as the timetable or the chronological ordering) directly relevant to our evaluation of modern scientific theories concerning the formative history of the universe, the solar system, the Earth, and the living creatures that move on its surface? Must the seven "days" of Genesis 1 be placed somewhere on the calendar of human historiography? Is God's work of creating categorically distinct from his providential work of sustaining and governing? According to the Bible, what are the characteristic features of the world that is God's created kingdom and in which mankind serves as God's image-bearing steward? And what implications do the answers to all of the above questions have for our reflections on natural science, Christian theology, and their interrelationships?

Drawing deeply from the well of biblical scholarship, John H. Stek deals at length with these and other important questions and offers the Christian reader answers that are both faithful to Scripture and shaped by the rich resources of the Reformed heritage. This chapter should not be read hastily, nor should its careful attention to textual detail go unappreciated as together we seek to know "what says the Scripture?"

SINCE THE RISE of the "scientific age," many have had the wrenching experience (though some claim to have found it "liberating") that, while science has "opened up the doors of the world" to them, it has "closed up the heavens forever."[1] That such might be true for someone catechized in one or another of the "primitive" religions is not surprising, but there is a sad irony in such an experience when reported by one early schooled in the Jewish or Christian faith. For the biblical view of God, humanity, and world has contributed much to the development of modern science. And the Bible itself affirms humanity's pursuit of knowledge of the world as a primary component in its God-given vocation.

Among those who acknowledge that vocation, one finds today, however, many who believe with a "stubborn faith" that the very biblical text that charters that vocation places certain rather specific restraints upon it. Though close examination of the Earth, the solar system, and the outer reaches of the cosmos has brought to light great masses of interlocking data that point to a specific history of development of the present physical universe spanning vast ages, they remain unyielding in their belief that the very word of God that calls mankind to the pursuit of knowledge "reveals" an origin of the creation that precludes such a formative history and such spans of time. As they read the Bible, God's "creating" by his word means instantaneous and *ex nihilo* calling into existence—both time and process are excluded. For them the narrative of creation in seven "days" provides God's own clear and straightforward chronicle of his specific creation acts distributed over a 144-hour period not many thousands of years ago. Similarly, the Deluge story "clearly"

1. The words are those of an Israeli teacher who early in this century spent her youth in Zefat, a town in the hills of Galilee where orthodox Jewish traditions had been long cherished. Priit J. Vesilind, "Israel: Searching for the Center," *National Geographic,* July 1985, pp. 2-5.

relates an account of a short-term global event with massive geological consequences. Hence all indications of "normal" processes at work through unimaginable spans of time are viewed as pure illusion. The "true believer" rejects the widely accepted "scientific" account of the Earth's formation and the progressive development of the cosmos as "unbiblical," the product of an enterprise that has miscarried because those pursuing it have methodologically excluded the one infallible source of knowledge about the origin and history of the physical universe.

In this chapter we seek to address these issues. Our question is that of Paul in Romans 4:3: "What says the Scripture?"

FOR ALMOST TWO THOUSAND YEARS Christians have pored over the biblical texts in an earnest effort to understand them. The greatest minds of the church have spent themselves in this consecrated endeavor. Not least among their concerns has been what the Bible teaches about creation. For this they turned especially to Genesis 1:1–2:3, and studies of the "hexaemeron" loom large among the writings they have left us.

Inevitably interpretations have been massively influenced by the philosophical and theological preoccupations of each age, the generally accepted world picture current in each generation, and other more subtle cultural factors. But until relatively recent times, the whole tradition of interpretation labored under the critical disadvantage of knowing little about the ancient world in which the Hebrew Scriptures were written. Lacking such knowledge, interpreters were left largely to speculation about it, guided only by what could be gleaned from the Old Testament itself, by later rabbinic traditions, and by what had been preserved from ancient Greek writings. Not until the eighteenth century did more information about that ancient world begin to come to light, and only in the twentieth century has it attained such scope as to yield some sound results.

Today interpretation of the Old Testament (not least Genesis) is undertaken on a new basis. The world out of which God called the patriarchs and within which ancient Israel emerged and walked with God (in both obedience and disobedience) has to a great extent been brought into the light of history: its peoples, its languages, its literature, its cultures, its politics, its religions, its "wisdom" (what it took to be knowledge about the world). As never before we are now aware of the relative lateness of Israel's appearance on the scene in the Near East. We can on the one hand perceive the degree of her cultural affinities with the peoples about her and on the other hand grasp more fully the scope and depth of the spiritual issues involved in God's wrestling for the soul of his people. The discovery and decipherment of whole libraries of ancient documents allow us for the first time to appreciate the range, sophistication, and subtlety of ancient literature: the variety, fluidity, and significance of its many genres (types); its creative and imaginative use of language; the protean forms of thought it embodies.

One engages in no cheap denigration of earlier interpretive efforts to judge that their results were in many respects inadequate and sometimes misguided. The best among them have achieved much that is valid and have bequeathed to us invaluable insights. But lacking adequate knowledge of the ancient context and bringing with them (all unconsciously) too much from their own, earlier interpreters often read into the biblical texts assumptions and issues alien to the biblical authors, resulting in distortions. The same dangers remain for us, of course. Those who simply take the Bible in hand and read it (in this or that translation) with an implicit faith that here God speaks to them directly in the idioms and forms of discourse current today are especially vulnerable. The inspiration of Scripture did not dehistoricize Scripture. However God's Spirit "moved" ("carried along," 2 Pet. 1:21, NIV) the biblical authors, they addressed their contemporaries concerning issues that were at stake in their day and in the language, conceptual modes, and literary forms familiar to their original readers. These facts must be taken into account if we would "correctly handle the word of truth" (2 Tim. 2:15).

Aware that many pitfalls beset the biblical interpreter, we shall attempt to lay out briefly our understanding of what the Bible says about creation and how this bears on the matters raised in our study. Because of space limitations and in view of its central importance, we shall concentrate mainly on Genesis 1:1–2:3. Our indebtedness to the history of interpretation will be evident, though a review of that history cannot be undertaken here. More obviously, perhaps, we shall seek to bring to the task what has been learned about ancient Israel's place in the long history of mankind and the cultural and religious environment that constituted her world.

For our purposes we focus on five related matters:

1. The biblical concept of "creation"
2. The nature and purpose of Genesis 1:1–2:3
3. The creation decrees and providence
4. Creation as divine kingdom; man as God's royal steward
5. Contemporary scientific cosmogony and the biblical doctrine of creation.

THE BIBLICAL CONCEPT OF CREATION

Biblical Language for Creating

What does it mean when the Bible says that God "creates"?[2]

bara'

The biblical language of creation is diverse. Most distinctive is the verb *bara'*. Though used (in this sense) in the Bible exclusively of the acts of God, it is highly unlikely that the basic Semitic root was originally coined as a specifically theological term. Israel's religious language was not fashioned in heaven but was drawn from everyday discourse: swords were made to serve as theological plow points and spears as theological pruning hooks. Inquiry into the context of human activity from which *bara'* was borrowed is hindered, however, by lack of adequate data.[3]

In biblical usage it obviously expresses the idea "to bring something into being," to bring into being some specific reality that had not existed before.[4] However, it does not *as such* signify giving existence to something that has never before existed *in kind* (e.g., a future human generation, Ps. 102:18; cf. 104:30 [cf. Isa. 42:5]; Isa. 45:7-8; Amos 4:13). Nor

2. We must, of course, be clear on what is meant in the English language by the word *create*. Common usage indicates that when a person is the subject (we also say that "the wind created a terrible dust storm"), it ordinarily expresses the thought that the subject has conceived of some new thing, willed to bring it into being, and then produced it. "Creating" can be inadvertent, as when we say that "his policies created great confusion in the enterprise," but most commonly "creating" involves deliberately bringing into being something that had been consciously conceived; we assume that this conception, will, and effectuation are all to be ascribed to the "creator"—as when we say, "He created a new machine" or a "new technique" or a "new literary form." The word as such says nothing about the means or manner or time involved or what, if anything, was utilized in the production. It only affirms that the subject bears the sole credit (or blame) for bringing some new thing into existence.

That, too (and more), is what is affirmed in the classic Christian confession: I believe in God, the Father Almighty, Creator of heaven and earth *(Credo in Deum Patrem omnipotentem Creatorem caeli et terrae)*. The final phrase, *Creatorem caeli et terrae*, though present in the confessions of Irenaeus and Tertullian, did not come in the Western Creed until well into the seventh century.

3. See C. Stuhlmueller, *Creative Redemption in Deutero-Isaiah*, Analecta Biblica 43 (Rome: Biblical Institute Press, 1970), pp. 209-13 (esp. nn. 674, 687); Karl-Heinz Bernhardt, *"bara',"* *Theological Dictionary of the Old Testament*, vol. 2, ed. G. J. Botterweck and H. Ringgren (Grand Rapids: Eerdmans, 1975), p. 245; and Claus Westermann, *Genesis 1–11* (Minneapolis: Augsburg, 1984), pp. 98-100.

4. Hence the frequent contextual reference to "newness": Ps. 51:10; 104:30; Isa. 48:6-7; 65:17; Jer. 31:22; cf. Exod. 34:10; Num. 16:30; Deut. 4:32.

is *bara'* used only of physical entities. In the language of the Old Testament, God "creates" historical events that effect judgment or redemption (Exod. 34:10; Num. 16:30; Isa. 48:7); he "creates" conditions that can be characterized as "darkness" and "evil" (Isa. 45:7) or as "righteousness" (an abstract noun indicative of a state of having been saved or delivered, Isa. 45:8); he "creates" a new quality of existence such as a transformed heart (a "pure heart," Ps. 51:10) or new conditions of a city ("Jerusalem to be a delight," Isa. 65:18); he "creates" praise on the lips of mourners ("fruit of lips," Isa. 57:19); he "creates" the smith (in his role as artificer) and the destroyer (in his role as devastator, Isa. 54:16). Since *bara'* (as "create") occurs in the Old Testament only with God as its subject, it seems to denote specifically (in biblical usage at least) a *divine* bringing into being. It views an existing reality and declares of it that God has by an incomparably divine act brought it into being—much as the signature on a painting announces that it was conceived, willed, and executed by the one whose signature it bears.

The frequent observation that God is never said explicitly to "create" *(bara')* "with" or "from" some prior existent is true enough, but its significance for determining the specific semantic value of *bara'* has been exaggerated. God's "creating" a "pure heart" (Ps. 51:10) or "Jerusalem a delight" (Isa. 65:18) or a new generation of people (Ps. 102:18) or of animals (Ps. 104:30; cf. Neh. 9:6), or his "creating" the smith or the destroyer (Isa. 54:16) hardly depicts a divine act that does not involve an action upon some already existent being. And alternative language for "creating" does refer to materials utilized. Genesis 2:7, for instance, tells us that God fashioned *(yaṣar)* man "from the dust of the ground" (cf. v. 19), and in v. 22 we read that God "built up [*banah*] the rib he had taken from the man into a woman." Still, the semantic limits of *bara'* seem to be such that it remains silent regarding that aspect of the action, focusing instead on the newness of the object created and the incomparable divine action by which it came into being.

Language Associated with *bara'*

References to *yaṣar* and *banah* lead to a consideration of Old Testament language closely associated with *bara'*. Just as God is called "the Creator of the ends of the earth" *(bara'*, ptc.: Isa. 40:28; cf. 42:5; 45:18), "the Creator of Israel" *(bara'*, ptc.: Isa. 43:15; cf. 43:1), "the Creator" of individual persons *(bara'*, ptc.: Eccl. 12:1), so he is called "the Maker of all things" *('asah*, impf.: Eccl. 11:5), Israel's "Maker" *('asah*, ptc.: Ps. 95:6; 149:2; Isa. 51:13; 54:5; Hos. 8:14), and the "Maker" of individuals *('asah*, ptc.: Job

4:17; 32:22; 35:10; Prov. 14:31; 17:5; Isa. 17:7).[5] In this series, the verb *'asah* is interchanged with *bara'*—as it is in Genesis 1:1–2:3, where *bara'* occurs in 1:1, 21, 25, 27; 2:3 and *'asah* in 1:7, 16, 25, 26, 31; 2:2, 3. The specific choice of words here may sometimes be deliberate, but it can be of little consequence that God is said to "make" rather than "create" the expanse that divides the waters above from the waters below (v. 7) and the heavenly bodies (v. 16), or that he is said on the one hand to "create" the forms of life that fill the waters and the air (v. 21) but on the other hand to "make" the animals that live on land (v. 25). Moreover, both verbs are used in conjunction with the creation of humankind (vv. 26-27). And in the closing summation of the account we read of "all that he had made" (1:31) and "all his work which God created" (2:3). Significantly, this final clause of the account combines both words in a tight syntactical construction: "for in it he rested from all his work which God created by making."[6]

As a matter of fact, the Hebrews employed a whole cluster of words to speak of God's creative acts; their ability to express the *concept* was not dependent on the availability of *bara'*. Of the Israelites who would experience God's future redemption, Isaiah wrote,

> Everyone who is called by my name,
> whom I have created *(bara')* for my glory,
> whom I have formed *(yaṣar)* and made *('asah)*. (Isa. 43:7)

Similarly he writes,

5. The familiar phrase "Maker *('asah)* of heaven and earth" (Ps. 115:15; 121:2; 124:8; 134:3; 146:6; cf. Gen. 2:4; Exod. 20:11; 2 Kings 19:15; 2 Chr. 2:12; Isa. 37:16) does not have a direct parallel with the verb *bara'*. In fact, the phrase "heaven and earth" as the object of *bara'* occurs only in Gen. 1:1 (cf. Isa. 65:17, "a new heaven and a new earth"). In Gen. 14:19, 22 the verb is *qanah*.

6. *Bara' . . . la 'asoth.* To "he created" *(bara')* has been added an infinitival phrase of the verb *'asah* in a construction that expresses further specification. While the precise force of the construction here is disputed, it brings together the two key verbs of the account. The immediate occasion for its use may arise from the presence of the noun *mela'kah*, which, like the English word "work," can refer either to the *activity* required to produce something (its most common usage), which can be "ceased," or to the product of the activity, which can be "created." In contexts of making something (such as the tabernacle and the temple and their furnishings), these two senses stand side by side (cf., for example, Exod. 39:43 and 40:33; see also 1 Kings 7:14, 51; 2 Chr. 4:11; 5:1). In Hebrew idiom, however, *mela'kah* is normally joined with the verb *'asah*—one "does work," as in Gen. 2:2: "And on the seventh day God was finished with all his work *(mela'kah)* which he had done *('asah)* and on the seventh day he ceased all his work *(mela'kah)* which he had done *('asah)*." Only in Gen. 2:3 does *(mela'kah)* occur with *bara'*. Hence the *sense* of the final clause may be more closely captured by paraphrasing it, "for in it God ceased all the work which he had been doing in creating."

> I form *(yaṣar)* the light and create *(bara')* darkness.
> I make *('asah)* prosperity and create *(bara')* disaster. (Isa. 45:7)

Particularly striking is his piling up of associated terms in 45:18:

> He who created *(bara')* the heavens,
> he is God;
> he who fashioned *(yaṣar)* the earth and made *('asah)* it,
> he established *(kun)* it.
> He did not create *(bara')* it formless;[7]
> to be a habitation he fashioned *(yaṣar)* it.

Similar combinations can be found in Deut. 32:6; Isa. 22:11; 27:11; 40:26; 43:1; 45:11-12; Jer. 10:12-16 (51:15-19); 33:2; and Amos 4:13.

The term *yaṣar,* to "fashion," occurs frequently (in addition to the instances noted above, see Ps. 33:15; 74:17; 94:9; 104:26; Isa. 43:21; 44:21, 24; Jer. 10:16; 51:19; Amos 7:1; Zech. 12:1). Sometimes the imagery of shaping as a potter is explicit (Isa. 29:16; 45:9; 64:8), but at other times such overtly physical imagery may well be left behind, as when God is said to "fashion" hearts (Ps. 33:15), summer and winter (Ps. 74:17), light (Isa. 45:7), disaster (Jer. 18:11), the events that shaped Israel's history (Isa. 22:11), or "the spirit of man within him" (Zech. 12:1).

Often God is said to have "established" the earth (*kun,* Ps. 24:2; 119:90; Isa. 45:18; Jer. 33:2; cf. Ps. 93:1; 96:10; Jer. 10:12 [51:15]—also the Moon and stars [Ps. 8:3], the heavens [Prov. 3:19; 8:27], the mountains [Ps. 65:6], and even "me" [Ps. 119:73])—as a throne is "established" (2 Sam. 7:13, 16; 1 Kings 2:45; Ps. 9:7; 93:2; 103:19; Isa. 16:5) or a kingdom (2 Sam. 7:12; 1 Kings 2:12, 46) or a house (Judg. 16:26, 29) or the temple mount (Isa. 2:2; Mic. 4:1) or a city (Ps. 48:8; 107:36).

God is also said to have "founded" the earth (*yasad,* Job 38:4; Ps. 24:2; 78:69; 89:11; 102:25; 104:5; Prov. 3:19; Isa. 48:13; 51:13, 16; Zech. 12:1)—as a temple is "founded" (1 Kings 6:37; Ezra 3:6, 10, 12; Hag. 2:18; Zech. 4:9; 8:9) or a city (Josh. 6:26; 1 Kings 16:34; Isa. 14:32).

Three times we read that God "spreads out" the earth (*raqa',* Ps. 136:6; Isa. 42:5; 44:24); the root being used in these instances elsewhere refers to trampling with the feet (2 Sam. 22:43; Ezek. 6:11; 25:6) or pounding out plates of metal (Exod. 39:3; Num. 16:39; Jer. 10:9; cf. Isa. 40:19).

As for the heavens, God "spreads them out" *(naṭah,* Job 9:8; Isa. 42:5; 44:24; 45:12; 51:13; Jer. 10:12 [51:15]; Zech. 12:1—once *raqa',* Job

7. Hebrew *tohu;* cf. Gen. 1:2.

37:18; once *ṭapaḥ,* Isa. 48:13) like a tent (Ps. 104:2; Isa. 40:22). But, as we have seen, he also "establishes" them *(kun,* Prov. 3:19; 8:27) and he "plants" them *(naṭaʿ,* Isa. 51:16).

Although their use in "creation" contexts is rare, note must also be taken of another word for "make" (*paʿal,* Job. 36:3; Prov. 16:4)[8] and of the expression "to build (up)" (*banah,* Gen. 2:22).

Finally, there is *qanah,* a much discussed root occurring occasionally in what appear to be "creation" statements (Gen. 14:19, 22; Deut. 32:6; Ps. 139:13; Prov. 8:22—for the noun, *qinyan,* see Ps. 104:24). Many biblical scholars judge that the root represents two nonsynonymous homographs, one meaning "to acquire, possess" (by means of purchase or otherwise), the other "to bring forth" (cf. Gen. 4:1, "I have brought forth [*qanah*] a man with [the help of?] the LORD"). The former is abundantly attested; the latter is still debated, but the "creation" contexts strongly suggest it.[9]

Much of this language is that of the poets; nonetheless, it makes abundantly clear that the biblical writers could speak in a rich variety of ways of God's bringing things into being and that *bara',* while the most specialized term, was closely associated with a variety of expressions that could be used virtually synonymously with it.

It must also be noted that whereas the Christian doctrine of creation has tended to focus almost exclusively on the *origin* of the creation *in the beginning,* that limitation does not apply to Old Testament creation language. In the speech of the Old Testament authors, whatever exists *now* and whatever *will come into existence* in the creaturely realm has been or will have been "created" by God. He is not only the Creator of the original state of affairs but also of all present and future realities. If the wind is blowing, he has created it (Amos 4:13); of each new generation of life it can be said that he created it (a "future generation" is "a people not yet created," Ps. 102:18; cf. 104:30); each human being has been created by him (Eccl. 12:1; has been "fashioned in the womb" by him, Isa. 44:2, 24; 49:5; Jer. 1:5; cf. Job 4:17; 10:8-12; 14:15; 31:15; 32:22; 33:4; 35:10; 36:3; 37:7; 40:15; Ps. 119:73; 139:13-16; Prov. 22:2; Isa. 17:7);

8. See also Exod. 15:17. The objects of this verb include arrows (Ps. 7:13), pit (Ps. 7:15), idol (Isa. 44:12, 15), iniquity (often), deceit (Hos. 7:1), righteousness (Ps. 15:2), and salvation (Ps. 74:12). The verb occurs in parallel with *kun* (Exod. 15:17; Ps. 7:13), *yaṣar* (Isa. 44:12), and *ʿasah* (Isa. 44:15). For the noun (*poʿal:* a thing made), see Isa. 45:9, 11.

9. For fuller discussions of this root, see Ernst Jenni and Claus Westermann, *Theologisches Handwörterbuch zum Alten Testament,* vol. 2 (München: Chr. Kaiser Verlag, 1979); *Theological Wordbook of the Old Testament,* vol. 2, ed. R. Laird Harris (Chicago: Moody Press, 1980); and B. Vawter, "Prov. 8:22: Wisdom and Creation," *Journal of Biblical Literature* 99 (1980): 205-16.

wherever there are conditions of distress, he is the Creator of them (Isa. 45:7; cf. Hab. 1:5); he is the Creator of each new generation of Israel (Ps. 100:3; 149:2; Isa. 27:11; 45:11; 46:4; 51:13; 54:5; Mal. 2:10); he is the "Maker" of all nations (Ps. 86:9); he is the Creator of the promised state of redemption (Isa. 4:5; 41:20; 45:8); and he is the Creator of the "new heavens and the new earth" (Isa. 65:17-18). Indeed, he is the "Maker of all things" (Eccl. 11:5; Jer. 10:16 [51:19]; cf. Ps. 104:24; 145:9, 10, 13, 17; Eccl. 3:11; Isa. 44:24). So when the Israelites confessed the LORD as "Creator *(qanah)* of heaven and earth" or "Maker *('asah)* of heaven and earth," it was of the very heavens and the very earth that lay before their eyes that they spoke.

Consonant with this, God's action of creating, preserving, and governing are not as sharply distinguished in Scripture as in Christian theological formulations. Note especially the words of Isaiah:

> Lift up your eyes and look to the heavens:
> Who has created all these—
> he who brings out the starry host one by one,
> and calls each by name?
> Because of his great power and mighty strength,
> not one of them is missing. (40:26; cf. Ps. 147:4)

> . . . he who is the Creator of the heavens and who
> stretches them out,
> who spreads out the earth and all that proceeds from it,
> who gives breath to the people upon it,
> and life-breath to what moves on it. . . .
> (42:5; cf. Ps. 104:30; Eccl. 12:7; Zech. 12:1)

Strikingly, the means by which God is said to have "created" *(bara')* may involve what from another perspective can be viewed as "normal" providential processes, such as his "creating" individuals (Eccl. 12:1; Ezek. 21:30; 28:13, 15; Mal. 2:10; cf. Ps. 89:47) or subsequent generations of people (Ps. 102:18; cf. 22:31) or animals (Ps. 104:30), or his "creating" the wind (Amos 4:13). Similarly, by his sovereign working in history, he "creates" Israel (Isa. 43:1, 7, 15), redemption (Isa. 41:20; 45:8; cf. 48:7), praise on lips (Isa. 57:19), darkness and disaster (Isa. 45:7), and the smith and the destroyer (Isa. 54:16). Apparently, whatever the means or manner and in whatever time God had brought something into being, it could, in the language of the biblical writers, be said that God had "created" *(bara')* it. Israel, it seems, had no specialized *technical term* for speaking specifically and exclusively of primeval creation or of instantaneous creation or of creation from nothing *(creatio ex nihilo/in nihilum)*.

The main conclusion to be drawn from this material is that while *bara'* had evidently come to be a specialized term in that it was used exclusively of God and refers to bringing something "new" into being, *it is silent as to the utilization of pre-existent materials or the time* (whether at the beginning of time or in the midst of time, whether instantaneously or over a period of time) *or the means involved.* In biblical language, *bara'* affirms of some existent reality *only that God conceived, willed, and effected it.*

The Manner of God's "Creating"

To this point the discussion has been limited in scope. We have addressed only the question of the vocabulary the Old Testament authors used to speak of "creating." As noted, that vocabulary was rich and varied, but it also had certain limits. It was abundantly adequate, however, to affirm that God sovereignly brought all that is into being.[10]

By Providential Means

When we ask *how* God "creates," we need to recall again that he is said to "create" individuals (Eccl. 12:1; Ezek. 21:30; 28:13, 15; Mal. 2:10; cf. Ps. 89:47), subsequent generations of people (Ps. 102:18; cf. Ps. 22:31) and of

10. In the Septuagint (a pre-Christian Greek translation of the Hebrew scriptures), we find that *poieo* ("to do, make") or *ktizo* ("to create, found") are generally used to translate *bara'* and *'asah;* the term *plasso* ("to fashion") is used to translate *yaṣar,* and the term *themelioo* ("to found, establish") is used translate *yasad.* In the Pentateuch, *poieo* is used to translate for both *bara'* and *'asah* in creation contexts, except in Gen. 2:3 (where *archo,* "to begin," is used) and Deut. 4:32 (where *ktizo* is used). In Gen. 14:19, 22, *ktizo* renders *qanah* (as in Prov. 8:22). In the Hagiographa, *ktizo* is regularly used to translate *bara'.* In the prophets, *poieo* is found in Isa. 42:5; 43:1; 45:7, 18; 65:18; *ktizo* is found in Isa. 45:7, 8; 54:16; Jer. 31:22; Ezek. 28:13, 15; Amos 4:13; Mal. 2:10; *katadeiknumi* ("to introduce something new") is found in Isa. 40:26, 41:20; 43:15; *kataskeuazo* ("to construct, build") is found in Isa. 40:28; 43:1; *histemi;* ("to set up, raise") is found in Isa. 65:17; and *gennao* ("to produce, create, call into existence") is found in Ezek. 21:30. At the time of the translation, *ktizo* was commonly used with the sense "to found, establish" a city (especially), temple, grove, theater, festival, or game. In such contexts it referred specifically to the basic intellectual and volitional act by which these came into being, in distinction from the actual execution. In post-Alexander Hellenism, the founding of a city was viewed in a special way as the work of a ruler, and a ruler was considered to embody some of the elements of divinity. Significantly, the Septuagintal translators did not employ the term *demiourgeo,* which denoted the technical and manual process of execution. Apparently, *ktizo* took on certain specialized connotations that recommended it for use in translating *bara'* subsequent to the translation of the Pentateuch. In the New Testament, the most common word for "to create" and "creature, creation" is *ktizo* and its derivatives. Far less frequently one finds *poieo* and its derivatives, and even less frequently *plasso.*

animals (Ps. 104:30), and the wind (Amos 4:13), and that in such instances his bringing into being can be viewed from another perspective—namely, that of providential processes. This is, of course, to employ a theological concept not found in the biblical literature.[11] The biblical authors spoke more concretely. Of the sequence of animal generations Psalm 104:29-30 says,

> You take away their breath, they expire,
> and to the dust they return.
> You send forth your Spirit, they are created,
> and the surface of the ground is renewed.

Concerning the sequence of human generations Psalm 102:18 says,

> Let this be written for the coming generation,
> and let the people to be created praise the LORD.

This is evidently a variation on the language of Psalm 22:30-31:

> Posterity will serve him;
> coming generations will be told about the Lord,
> and they will proclaim his righteousness to a people
> to be born. . . .

Reference to the womb is explicit in Job 31:15:

> Did not he who made me in the womb make them?
> Did not the One form us within the womb?

Similarly in Psalm 139:13, 15:

> For you created *(qanah)* my inmost being:

11. The nearest the biblical authors come to expressing the theological concept of "providence" appears to be found in Job 10:12:

> You have given me life and love
> and your *attentive care (pequddah)* has preserved my (life-)spirit.

Elsewhere what is encompassed by "providence" finds expression in a variety of terms of more limited semantic range, including the following: to bless *(barak)* or curse *('arar, qalal* [Pi.]); to preserve, guard, watch over *(šamar, naṣar, ḥazaq* [Hiph.], *kul* [Pilpel]); to lead, guide *(nahag, nahal, naḥah);* to remember *(zakar);* to "visit" caringly *(paqad).* Such is the language used with respect to living creatures, especially humans. Another range is employed in speaking of the cosmic order of the creation: having founded *(yasad),* fashioned *(yaṣar),* and established *(kun)* the world by his sovereign word *(dabar)* in accordance with his decrees *(ḥuqqim)* for them, and having covenanted to maintain them (Jer. 33:20, 25), he sovereignly governs them in faithfulness *('emet, 'emunah),* in steadfast love *(ḥesed),* in righteousness *(ṣedeq),* and in justice *(mišpaṭ)* (cf. Ps. 33:4-5; 36:5-6; 57:10; 89:2, 5, 8; 119:89, 90; 145:17).

> you wove me together in my mother's womb. . . .
> I was woven together (another verb). . . .[12]

After speaking of the LORD as Israel's "Creator" (43:1, 15), Isaiah says of him that

> he . . . made (*'asah*) you (Jacob), . . . formed (*yaṣar*) you in the womb. (44:2; cf. vv. 21, 24; see also 49:5.)[13]

There is no warrant for supposing that God's "creating" of new generations of life in the womb was thought to be unrelated to the sexual union of male and female. Already in Genesis 4:1 we read: "Adam lay with his wife Eve, and she conceived and gave birth." This type of birth report is actually conventional in the Old Testament literature. Even more explicitly to the point is the statement that Onan, "whenever he lay with his brother's wife, . . . spilled his semen on the ground to keep from producing offspring for his brother" (Gen. 38:9). Moreover, when Jacob attempted by magical means to control the color of the offspring of Laban's flocks, he was shown in a dream "that the male goats mating with the flock were streaked, speckled and spotted" (Gen. 31:10).

The fructifying of the womb by which new generations of humans and animals are produced is frequently ascribed by the biblical writers to God's blessing. Here we must recall God's primeval benediction on birds and marine life (Gen. 1:22) and on humanity (v. 28): "Be fruitful and multiply and fill the water in the seas and let birds multiply on the earth . . . and fill the earth." Accordingly, the LORD promised to bless Sarah so that she would have a son (Gen. 17:16), to bless Ishmael so that he would "be fruitful and multiply very much" (v. 20), to bless Isaac so that his descendants would be "multiplied" (Gen. 26:24), and to bless Israel so that it would "multiply" (Deut. 7:13). From this perspective, God "creates" new generations of life by effecting his blessing.[14]

The Old Testament also employs a richly varied language in describing God's work in producing the wind. In the hymnic language

12. Or "embroidered"—if the root *raqam* is not a corruption by metathesis of *qaram*, in which case the sense would be "covered with skin" (cf. Ezek. 37:6).

13. In biblical perspective, only God can "give" children (Gen. 30:1-2; cf. 1 Sam. 1:11, 27; 2:20), which he does by "opening the womb" (Gen. 29:31; 30:22). Similarly, only God can withhold children (Gen. 16:2; 30:2), which he does by "closing the womb" (Gen. 20:18; 1 Sam. 1:5, 6; Isa. 66:9) or giving a "bereaving womb" (Hos. 9:14). He is also the One who "brings forth" from the womb (Job 10:18; cf. Ps. 22:9; 71:6—but the text and/or precise meaning of the verbs in the last two texts are in dispute).

14. C. Westermann, "The Blessing of God and Creation," in *Elements of Old Testament Theology* (Atlanta: John Knox Press, 1982), see especially pp. 102-14.

of Amos 4:13, God "creates" the wind. In Psalm 135:7 and Jeremiah 10:13 (51:16), God "causes it to go forth from his storehouses." Elsewhere it is said that God "conducts" it (*nahag:* Exod. 10:13; Ps. 78:26) or "causes it to set out" (*nasaʿ*, as from an encampment: Ps. 78:26; cf. Num. 11:31—language used also of a community of people or of a flock, Ps. 78:52). Ezekiel says that God will cause a violent wind to "break out" (or "burst forth," *baqaʿ*, 13:13; cf. 13:11). Jonah says that God "hurled" *(ṭul)* a great wind on the sea (1:4) and that he "appointed" *(manah)* a hot east wind (4:8). Hosea says simply that "An east wind will go forth from the LORD" (13:15), and the author of Genesis says that God "caused a wind to pass over" the earth to dry up the water of the flood (8:1). In an expression of somewhat uncertain meaning, Job 28:25 speaks (in a "creation" context) of God "making *(ʿasah)* weight for the wind" (giving force to it?). Most vivid of all is the language of Psalm 104:4:

> He makes *(ʿasah)* the winds his messengers,
> lightning his ministering ones.

And in Psalm 148:8 the "stormy winds that do his bidding" are listed among the creatures of the earthly realm (cf. v. 5) that are exhorted to praise the LORD.

All these various expressions ascribe the origin of the wind to God without indication of intermediate causes.[15] No doubt the ancients had little understanding of the "natural" causes of winds, and so it is not strange that they would have assigned their origin to God. But we have no evidence that the biblical authors viewed every breeze and whirlwind as a "wonder" *(nipla't);* winds were as much a part of man's everyday environment as the mountains/hills (with which they are linked in Amos 4:13; cf. Ps. 148:8-9). Nevertheless, it could be said, at least in the language of praise, that God "creates" them.

By His Word—By Divine Fiat

When the biblical writers wished to declare expressly how God brings into being and establishes the order of his creation, they most commonly said that he does so by his word (command)—that is, by divine fiat.[16]

> By the word of the LORD were the heavens made,
> their starry host by the breath of his mouth. . . .

15. Hence also the wonder evoked in the apostles when Jesus stilled the wind and the waves with a word (Matt. 8:27; Luke 8:25).

16. The classic expression comes from Latin *fieri,* 3rd m.s., "Let it be done/become."

For he spoke, and it (the world) came to be;
he commanded, and it stood forth. (Ps. 33:6, 9)

Let them (the whole celestial host) praise the name of the LORD,
for he commanded and they were created.
He set them in place forever and ever;
he gave a decree that will never pass away. (Ps. 148:5-6)

Indeed, my hand laid the foundations of the earth,
and my right hand spread out the heavens;
at my summoning them,
they stood forth together. (Isa. 48:13)

To much the same effect are the following:

Thus says the LORD,
who appoints the sun to shine by day,
decrees for the moon and stars
so that they shine by night,
who stirs up the sea so that its waves roar—
the LORD Sabaoth is his name:
"Only if these decrees vanish
from my sight," declares the Lord,
"will the descendants of Israel ever cease
to be a nation before me." (Jer. 31:35-36; cf. 33:25; Job 38:33)

It is I, who made the earth
and created mankind upon it.
My hand stretched out the heavens;
I commanded their starry hosts. (Isa. 45:12)

Who shut up the sea behind doors
when it burst forth from the womb, . . .
when I fixed limits for it
and set its doors and bars in place,
when I said, "Thus far you may come and no farther;
here is where your proud waves halt"?
(Job 38:8-11; cf. Ps. 104:9; Jer. 5:22)

He fills his hand with the lightning,
and commands it to strike its mark. . . .
He says to the snow, "Fall on the earth,"
and to the rain shower, "Be a mighty downpour." (Job 36:32; 37:6)

He established limits for the sea
so the waters can not transgress his command. (Prov. 8:29)

He sends his commands to the earth;
his word runs swiftly.
He spreads the snow like wool
and scatters the frost like ashes. . . .
He sends his word and melts them;
he stirs up his breezes, and the waters flow. (Ps. 147:15-18)

He calls for the waters of the sea
and pours them out over the face of the land.
(Amos 5:8; 9:6)

These examples are all gleaned from poetic materials. They depict the Creator as the heavenly Sovereign who effects his establishing and governing of the creation by his word of command (note the language: "decree" [*ḥoq*], "command(s)" [*ṣawah, peh, 'imrah, dabar*], "summon" [*qara'*]). The classic presentation of God's creative word is found, of course, in Genesis 1, where God's word of command brings into being (vv. 3, 6, 9, 11, 14, 20, 24), appoints creaturely functions and spheres of dominion (vv. 14-18, 26), assigns sustenance (vv. 29-39), and blesses (vv. 22, 28). Some interpreters suppose that all biblical talk of divine creation by word is a reflex of Genesis 1, but that is a questionable assumption. Be that as it may, it is relevant to observe that *the concept of divine creation by a word of command does not of itself assume instantaneous effectuation.* Just as with God's "commands" concerning rain, snow, frost, and melting (Job 37:6; Ps. 147:15-18) and his directing word concerning historical events, *the point of the biblical authors is the sovereignty and effectiveness of God's creative word and directive rule over the creaturely realm, with no reflection on the time or the mediating agents and/or processes involved.*

Perhaps this assertion needs further demonstration. When Elihu declares that God "commands" the lightning to strike its mark (Job 36:32) and says to the snow, "Be upon the earth," and to the rain, "Be a mighty downpour" (Job 37:6), he is declaring that lightning, snow, and rain, wherever they fall on the earth, come by divine fiat. The psalmist asserts the same thing in 147:15-18, adding to the meteorological catalogue frost, hail, cold winds, and balmy breezes. Yet this divine fiat involves neither instantaneous nor unmediated effectuation.

Similarly, we read that God "calls/summons" *(qara')* drought and famine (2 Kings 8:1; Ps. 105:16; Hag. 1:11), "calls (together)" kings (2 Kings 3:10, 13), summons his warriors to execute his judgment (Isa. 13:3; cf. Lam. 1:15; Ezek. 38:21), and calls his servant Cyrus to crush Babylon (Isa. 48:15). At his "command" *(ṣawah)* judges are raised up over his people (2 Sam. 7:11), the ravens and a Phoenician widow feed Elijah (1 Kings 17:4, 9), God's "holy ones" carry out his judgments (Isa. 13:3), desert creatures inhabit Edom (Isa. 34:16), the Babylonian army returns to Jerusalem (Jer. 34:22), Israel's neighbors become her foes (Lam. 1:17), God's purpose to punish Jerusalem is carried out (Lam. 1:17), the great and small houses of Israel are reduced to rubble (Amos 6:11), and Israel is "shaken" among all the nations (Amos 9:9). God also "commands" justice (Ps. 7:6), his love (Ps. 42:8), victory for his people (Ps. 44:4), deliverance (Ps. 71:3), and his blessing (Ps. 133:3). In fulfillment of God's command *(peh)* the nations overrun Judah (2 Kings 24:3), and both calamities and good things come (Lam. 3:38). God's "word" *(dabar)* effects healing (Ps. 107:20) and judgment (Isa. 9:8; Jer. 23:29)—every divine purpose for which he "sends" it (Isa. 55:11). Even his word placed on the lips of his messenger prophets is powerful to effect the divine purposes they announce: "Now, I have put my words in your mouth. See, today I appoint you over nations and kingdoms to uproot and tear down, to destroy and overthrow, to build up and to plant" (Jer. 1:9-10, cf. 5:14; Hos. 6:5). The word concerning his purposes that God commands his prophets to announce is once called his "decree" (Zech. 1:6).

In the language of the Old Testament, it can be said of all that God effects in the world (alike in "nature" and "history") that he does so by his word. Such language belongs to the pervasive depiction of the sovereign God as the heavenly King, the universal and absolute Monarch. He speaks, commands, summons, decrees, and it comes to pass—whatever the "natural" means utilized, whatever the "historical" forces and agents employed, whatever the span of time involved. His sovereign word is infallibly effective. *Whatever he does, he does by fiat.*

Against the background of this pervasive Old Testament language, it is evident that the focus of Psalm 33:9—

> He spoke, and it (the world) came to be;
> he commanded, and it stood forth.

—lies on the sovereign effectiveness of the divine word, not on the immediacy of its effectuation. This is put beyond doubt by the similar formula of Lamentations 3:37-38:

> Who can speak, and it comes to be,
> if the Lord has not commanded (it)?
> Is it not from the mouth of the Most High
> that calamities and good things proceed?

Only God's word has the absolute power to effect its purpose: both disaster and prosperity, by whatever means they might come, are from him—as God himself testifies:

> I form the light and create darkness;
> I bring prosperity and create disaster;
> I, the LORD, am the One who does all these things.
> (Isa. 45:7)

God creates by his sovereign and effective word. Creation is by divine fiat. But that assertion of itself is silent as to whether God's creative word achieved its purpose immediately (without mediating agency or process) *and instantaneously or mediately and over the course of time.*[17]

Creation ex nihilo

Our survey of the biblical data has shown that the Old Testament authors' talk of divine "creating" did not inherently imply creation *ex nihilo* or *in nihilum* ("out of nothing" or "in nothingness"). The question remains, however, whether or not the author of Genesis 1 conceived of God's original creation of the world as *ex nihilo.* The author of 2 Maccabees (early first century B.C.) clearly thought so, but, as far as is presently known, he is the first to have expressed something like this formula in writing.[18] The author of the epistle to the Hebrews appears to have had this understanding of the Genesis account when he wrote, "By faith we understand that the universe was formed at God's command, so that what is seen was not made out of what was visible" (11:3). The early church fathers also appear to have taken it for granted.

These interpretations of Genesis 1, however, all come well after the inroads of Hellenism with its Greek modes of thought and cosmic speculations. Many interpreters of the Old Testament have voiced doubts that earlier Israelites and their Semitic neighbors had attained the kind of abstract conceptualizing that the *ex nihilo* formula presupposes. Be that as it may—and what the limits of the Israelite powers of conception may have been is difficult to establish with any certainty—Walter

17. This being the case, the immediacy and instantaneousness of various miracles in the ministries of God's servants in the history of redemption is not really relevant.

18. In 2 Macc. 7:28 we read, "I beseech you, my child, to look at the heaven and the earth and see everything that is in them, and recognize that God did not make them out of things that existed" (or, "God made them out of things that did not exist"). Translation and alternative from RSV.

Eichrodt is no doubt close to the mark when he observes that the author's strict monotheism, the radical difference between his cosmogony and that of the current myths with their assumption of a primordial substrate from which the first generation of the gods emerges, his opening reference to an "absolute beginning for the creation," his emphasis on creation by divine word, and his choice of the verb *bara'* in the superscript (v. 1) all tend toward a concrete depiction of original creation for which anything short of *ex nihilo* in more abstract language would fail to do justice.[19] Nevertheless, although for us *ex nihilo* in its strictest sense certainly qualifies the absolute beginning of the creation, *it does not necessarily apply to all of God's acts that are viewed as "creative" by the Old Testament writers.*

We have gone into this much detail because the biblical language pertaining to divine creation has been so widely misconstrued. Contrary to claims often made, while the *verbs* employed affirm in richly varied ways divine effectuation—the idea that God with sovereign power and will has made and ordered all things—they do not specify *ex nihilo* or instantaneous creation or the absence of process or mediating agencies. Nor does the fact that God creates by fiat circumscribe more narrowly the means or manner of his creative acts, as commonly alleged. That the original creation was *ex nihilo* appears to be implied in the grand pronouncement with which Genesis begins, but that is a conclusion that can be drawn only from the sense of the full statement in its literary and historical contexts. The Bible rarely contains terms of the highly technical sort that are employed in philosophical or scientific (or even theological) essays. Their semantic values must in each case be discerned from close attention to context. Of all the biblical verbs used to refer to divine creation, *bara'* is clearly the most specialized, but even this term is silent on the *how* of God's creating, as we have seen.

THE NATURE AND PURPOSE OF GENESIS 1:1–2:3

No doctrine is more widely attested or assumed in the biblical literature or is more foundational for the biblical message than that of God's creation of "all things visible and invisible."[20] The classical passage, as

19. Eichrodt, *Theology of the Old Testament,* vol. 2 (Philadelphia: Westminster Press, 1967), pp. 101-6.

20. For new appreciation of this fact among Old Testament scholars, see H. H. Schmid, "Creation, Righteousness and Salvation," in *Creation in the Old Testament,* ed. B. W.

universally recognized, is Genesis 1:1–2:3, the words with which the biblical story begins.[21] Still, even here the Bible does not speak of God's creation of the world as a separate and isolated topic of interest. This marvelously lapidary, tightly controlled, and theologically rich account of God's first acts serves as prologue to the story of his dealings with mankind and with the "fathers" of Israel, which in turn serves as introduction to the Pentateuch.[22] While 2:4–3:24 initiates that story at the ultimate *historical* horizon (viz., the creation of mankind, their fall, and their expulsion from the Garden with its tree of life), 1:1–2:3 sets it within its fundamental theological, cosmological, and anthropological context. It proclaims that the God who has related himself to Israel through redemption and covenant is the Creator of the world and all that is in it, that the world and all that is in it are works of his hands and subject to his rule, and that men and women, as beings created "in his image," are his servants and royal stewards in the visible creation. Thus it supplies the fundamental view of God, humanity, and world within which alone the subsequent narrative makes sense. In this opening movement of the biblical symphony we hear most powerfully (rivaled only by the prologue to the Gospel of John) that the God who comes in redemption and blessing (but also in judgment and curse) is none other than the Creator of all that is, and that the ultimate power and purpose

---

Anderson (Philadelphia: Fortess Press, 1984), pp. 102-17; R. P. Knierim, "The Task of Old Testament Theology," *Horizons in Biblical Theology* 6 (1984): 25-57; and Roland E. Murphy, "Wisdom and Creation," *Journal of Biblical Literature* 104 (1985): 3-11.

21. The long-held view of critical scholars that Gen. 2:4a concludes this unit is again being effectively challenged, so that the traditional and more probable understanding of the unit's boundary is being rehabilitated. On this, see B. W. Anderson, "A Stylistic Study of the Priestly Creation Story," in *Canon and Authority,* ed. G. W. Coates and B. O. Long (Philadelphia: Fortress Press, 1977), pp. 148-62.

22. One may question certain implications in his words, but von Rad is surely right on the central issue when he says, "Genesis 1 is not an independent theological essay, but one component of a great dogmatic treatise" ("The Theological Problem of the Old Testament Doctrine of Creation," in *Creation in the Old Testament,* ed. B. W. Anderson, p. 60 [reprinted from *The Problem of the Hexateuch,* 1966]).

On the basic issue of the nature and function of what von Rad calls "a great dogmatic treatise" (the Pentateuch), M. G. Kline is no doubt closer to the truth: "If the Pentateuch is viewed as a unified corpus with God's covenant with the exodus generation of Israel as its nucleus, the narratives of Genesis and the first part of Exodus assume the character of a historical prologue tracing the covenant relationship to its historical roots in Yahweh's past dealings with the chosen people and their patriarchal ancestors" (*The Structure of Biblical Authority* [Grand Rapids: Eerdmans, 1972], p. 53; Kline wisely disclaims contending thereby that "The Pentateuch as such is a treaty in form," p. 53 n. 17).

On this understanding, Gen. 1:1–2:3 is prologue to the "historical prologue" of the Sinai covenant. Cf. M. G. Kline, "Genesis," in *The New Bible Commentary,* rev. ed. by D. Guthrie and J. A. Motyer (Grand Rapids: Eerdmans, 1970), p. 81.

at work in history is none other than the power and purpose at work in creation.[23]

Sources and Date

The source(s) and date of this account of creation have long been hotly debated. Since the classic formulation of the documentary hypothesis by Julius Wellhausen in the late nineteenth century, it has become a virtual article of faith among critical scholars that Genesis 1:1–2:4a is a piece of postexilic priestly (P) theology.[24] Even so, most have recognized that it was based on much earlier traditions.[25] Evangelicals have for the most part held to a Mosaic authorship, but they, too, have usually assumed that Moses utilized traditions.[26] Their further assumption that Moses had available to him traditions that reached back in unbroken lineage to the first human generation is, however, open to serious question. Data assembled by historical anthropologists, especially during the past century and a half, provide extremely strong warrant for the conclusion that human origins date to a dim and distant past and that the course of human history during those long ages was such that it is inconceivable that a core tradition could have been accumulated and perpetuated through which Israel's patriarchs could trace their history back to humankind's first pair.

On the basis of careful examination of scores of archaeological sites, investigators have concluded that communal living in permanent

23. "The reason why this chapter is at the beginning of the Bible," writes Westermann, "is so that all of God's subsequent actions—his dealings with humankind, the history of his people, the election and the covenant—may be seen against the broader canvas of his work in creation" (*Genesis 1–11*, p. 175).

24. The documentary hypothesis holds that the Pentateuch and possibly also Joshua and the early part of Judges were composed by stitching together four earlier documents (usually identified by the sigla J, E, D, and P), the earliest of which (J) dates from the tenth century B.C. and the latest of which (P) dates from the postexilic period. See J. Wellhausen, *Die Composition des Hexateuchs*, 1889.

25. Since H. Gunkel's *Schöpfung und Chaos in Urzeit und Endzeit* (1895), these earlier traditions have generally been supposed to be Mesopotamian myths reworked and demythologized. See the abridged translation of Gunkel's work under the title "The Influence of Babylonian Mythology upon the Biblical Creation Story," in *Creation in the Old Testment*, ed. B. W. Anderson, pp. 25-52.

26. Gleason Archer's view may be taken as representative of the evangelical position: "While materials which the author used for the composition of this book no doubt came to him from five or six centuries before his time, prior to Jacob's migration into Egypt, nevertheless Moses seems to have served as the Spirit-guided compiler and interpreter of the pre-existent material which had come to him from his forebearers in oral and written form" (*A Survey of Old Testament Introduction*, rev. ed. [Chicago: Moody Press, 1974], p. 179).

settlements, with agriculture and/or pastoralism as the main modes of subsistence, developed in the Near East only in the tenth to ninth millennia B.C. From such early agricultural villages, cities gradually emerged in southwest Asia after 8,000 B.C. Though none of the earliest cities (such as Jericho in the Jordan valley and Catal Hüyük in south-central Turkey) endured, the fourth millennium B.C. saw the rise of more complex city-states along the Tigris and Euphrates rivers, the first of which was Uruk (biblical Erech). By 3,000 B.C. something like city-states could be found along the Nile in Egypt, and the third millennium witnessed similar developments in the Indus River valley, most spectacularly the great city of Moenjo-Daro, three hundred miles north of Karachi in Pakistan. (Transition to permanent settlements and later to urban centers took place much later in the Americas; Australian aborigines were still hunter-gatherers when that continent was discovered by Europeans in the seventeenth century.) Prior to these cultural developments, small human bands made up of some three to six core families seasonally ranged limited geographical areas as hunter-gatherers. Evidence from thousands of sites shows that for tens of thousands of years such wandering bands were widely dispersed over the Eurasian-African landmass. By 30,000 B.C. they were present in Australia, and in the Americas by 12,000 B.C., if not much earlier.[27]

The assumption of earlier evangelical theologians that it makes little difference for our reading of the early chapters of Genesis whether the time between the first humans and Abraham is to be measured in thousands of years, tens of thousands of years, or even hundreds of

27. Although the literature dealing with the relevant data concerning this early human history is massive, including thousands of field reports, it is little known except by those involved in historical anthropology. Those not familiar with the data might begin with such global summaries as Grahame Clark, *World Prehistory in New Perspective,* 3rd ed. (New York: Cambridge University Press, 1977); Robert J. Wenke, *Patterns in Prehistory: Humankind's First Three Million Years,* 2nd ed. (New York, Oxford: Oxford University Press 1984); and the associated literature noted in these volumes. For some recent regional surveys, see Herbert Schutz, *The Prehistory of Germanic Europe* (New Haven: Yale University Press, 1983); Timothy Champion, Clive Gamble, Stephan Shennan, and Alasdair Whittle, *Prehistoric Europe* (New York: Academic Press, 1984); and Linda S. Cordell, *Prehistory of the Southwest,* New World Archaeological Series (New York: Academic Press, 1984). The careful reader will, of course, distinguish between the data cited and the interpretations placed on them.

Recognition of human antiquity did not come easily for a variety of theological and scientific reasons. One of the ironies of history is that it came among those of scientific reputation in Britain and Europe in 1859, the same year that Charles Darwin published *The Origin of Species.* It came, however—and this is important to note—quite independently of Darwin's evolutionary theory. The history of this revolutionary shift in views concerning the age of mankind has been carefully traced by Donald K. Grayson in *The Establishment of Human Antiquity* (New York: Academic Press, 1983).

thousands of years seems to have been based on the belief that somewhere in the Near East there must have existed a stable core human community from which all others were detached and dispersed and that that core community maintained a distinct identity through hundreds if not thousands of generations so that it was able to preserve a communal memory of its past, including remembrances of primeval events.[28] The weight of evidence from anthropological investigations makes that supposition unsupportable.

That being the case, we must now turn to the question of the genesis of Genesis 1:1–2:3. All things considered, the *earliest* likely date for some account of God's creation of the world in six days, culminating in a seventh day of rest, appears to be the time of the Israelite patriarchs.[29]

28. See, for example, B. B. Warfield, "The Antiquity and the Unity of the Human Race," *Princeton Theological Review* 9 (1911): 1-25. Warfield states without qualification that "The question of the antiquity of man has itself no theological significance. . . . The Bible does not [in its genealogies] assign a brief span to human history; that is done only by a particular mode of interpreting the Biblical data, which is found on examination to rest on no solid basis" (pp. 1, 2). He was convinced of this point especially by William Henry Green, "Primeval Chronology," *Bibliotheca Sacra* 47 (April 1890): 285-303. He concluded that "For aught we know instead of twenty generations and some two thousand years measuring the interval between creation and the birth of Abraham [the traditional chronology based on the genealogies in Gen. 5 and 11] two hundred generations, and something like twenty thousand years, or even two thousand generations and something like two hundred thousand years may have intervened" (p. 11). "The question of the antiquity of man," he said, "is accordingly a purely scientific one, in which the theologian as such has no concern" (p. 11).

In the article to which Warfield referred, Green repeated and developed an argument he first published in 1863 (*The Pentateuch Vindicated from the Aspersions of Bishop Colenso*). Quoting from that earlier publication, Green reaffirmed his view that "if the recently discovered indicators of the antiquity of man . . . shall, when carefully inspected and thoroughly weighed, demonstrate all that any have imagined they might demonstrate, what then? They will simply show that the popular chronology is based upon a wrong interpretation [of the Genesis genealogies], and that a select and partial register of ante-Abrahamic names has been mistaken for a complete one" ("Primeval Chronology," pp. 285-86).

The date of Green's original publication (1863) is significant in that it falls only four years after the scientific community had finally concluded that the artifactual data then in hand no longer allowed for any other conclusion than that humanity's antiquity was far older than had been believed.

Both Warfield and Green assumed, however, that the genealogies and associated narratives in Gen. 1–11 were based on faithfully transmitted traditions going back to the first humans. But at the end of the nineteenth century, prehistoric anthropology was only in its infancy. Massive subsequent findings by specialists working in the field during this century have rendered their assumption untenable.

29. A number of factors bear on this question: (a) Assuming the earliest possible date for Genesis in essentially its final form, the earliest *documentation* of the account comes from the time of Moses. (b) If this account of creation pervasively evokes the royal metaphor, it cannot predate the emergence of monarchies as a well-developed and long-established political form in the Near east. (c) If, as seems evident, this account of creation polemically engages the documented cosmogonies of the ancient Near East, it must

But even so early a date is problematic, since we have no indication that the patriarchs observed a weekly Sabbath. According to the biblical account, the Sabbath was first observed as a divine ordinance by Israel after their deliverance from Egypt (Exod. 16). Perhaps we can say no more than that the Sabbath ordinance that God first laid down for Israel after the exodus presupposed some account of divine creation that had a sabbatic structure (if only in so simple a form as that found in Exod. 20:11) and that Genesis 1:1–2:3 became the final canonical form of that account. In any event, a compelling body of evidence indicates that the Genesis account of creation cannot credibly be viewed as a tradition handed down through the generations from humanity's first parents to Israel's patriarchs but must instead be viewed as a word about creation composed no earlier than the second millennium B.C.

Conceptual Affinities with the Ancient Near East

That this account of creation reflects basic conceptual affinities with Near Eastern cosmologies of the second millennium is recognized by all who are familiar with the literature and the pictorial representations recovered from Israel's cultural environment.[30] In Egyptian iconography, common features of cosmic representation include the star-studded body of a female figure (the sky goddess Nut) arched over the flat earth (see p. 227).[31] Above her the barque of the Sun sails the heavenly ocean from horizon to horizon. A circle formed by (or two concentric circles having between them) a serpent with its tail in its mouth represents linkage between the ocean above the starry heaven and that below the earth, the subterranean ocean (the god Nun) which the Sun barque traverses at night to make its dawn appearance in the east. The earth itself (the god Geb) is sometimes portrayed symbolically as a round disk

---

necessarily be dated subsequent to the emergence and dispersal of those myths. (d) The sabbatic structure of this account must be related in some way to the establishment of the basic sabbatic structure of time in Israel's life. The earliest biblical reference to regular Sabbath observance occurs in connection with the gift and gathering of manna (Exod. 16), and the subsequent Sabbath stipulation of the Sinai covenant motivates observance of the weekly Sabbath by appeal to God's rest on the seventh day of creation, which he "therefore" blessed and consecrated (Exod. 20:8-11). Accordingly, later tradition held that God had made known to Israel "your holy Sabbath" (Neh. 9:14).

30. For examples of these pictorial representations, see especially Othmar Keel, "Conceptions of the Cosmos," in *The Symbolism of the Biblical World: Ancient Near Eastern Iconography and the Book of Psalms* (New York: Seabury Press, 1978), pp. 16-60.

31. Sometimes the sky is represented as a flat roof-like structure resting on mountains that rim the earth.

In this highly symbolic Egyptian representation of the world, the starry sky (the lady of heaven—the goddess Nut) arches over the reclining earth (the god Geb). Above her is the upper ocean in which the solar barque sails to carry the sun (the falcon-headed god Re) from the eastern horizon up to the zenith and then down to the western horizon. (Re is accompanied by the goddess Maat with her identifying feather; she is the daughter of Re, who is the source of world order.) Kneeling above the earth (reclining Geb) and holding up the sky (Nut) is the atmosphere god Shu; he holds in both hands the symbol of the breath of life. At the lower right is Osiris, the great god of the world of the dead. Not represented here is the subterranean ocean which the solar barque traverses at night to return the sun at dawn to the eastern horizon. That ocean is clearly shown in other symbolic representations. (From Othmar Keel, *The Symbolism of the Biblical World: Ancient Near Eastern Iconography and the Book of Psalms* [New York: Crossroad, 1985])

surrounded by water. Hence the Egyptian view of the cosmos can be represented by two circles: a vertical circle linking the ocean above with the ocean below and a horizontal circle of the earth surrounded by the waters of the lower sea. Thus the cosmos is like a sphere bounded by the linked upper and lower oceans. Enclosed within these oceanic bounds are the earth and the visible heavens.

The earth is also portrayed as a disk surrounded by ocean in a Babylonian map of the world. Iconographic representations among the Mesopotamians and peoples of related areas tended, however, to be more directly representative of mythic themes than of structural schemes. Nevertheless, it is evident that in Mesopotamia, too, the cosmos was conceived as made up of an ocean above the starry heavens, the visible celestial realm itself, the earth, and an ocean below. In *Enuma elish,* a major creation myth possibly dating from the time of Hammurabi (the end of the eighteenth century B.C.),[32] Marduk overcame Tiamat (the primeval saltwater ocean and one of the pair of primeval gods ancestral to all the other gods) and then

> He split her like a shellfish into two parts;
> Half of her he set up and ceiled it as sky,
> Pulled down the bar and posted guards.
> He bade them to allow not her waters to escape.[33]

Thereafter he took the measure of Apsu, the ocean below (presumably the other half of Tiamat) and made over it a firmament (canopy) in which he established the celestial bodies. Finally he formed the mountains (earth) on Tiamat's head. Thus he created "heaven and earth," situated between the ocean above and the ocean below.[34]

As is evident both in Egyptian cosmogonies and in *Enuma elish* the original substrate was conceived of as a watery mass, which was subsequently divided to form the upper and lower oceans. In the space between, "heaven and earth" were formed. We have seen how this was said to come about in *Enuma elish.* According to the dominant Egyptian cosmogony, the first cosmogonic development saw the emergence of the primeval hill out of Nun (primordial water). This primeval hill (of which the pyramids were a representation) was Atum, the first of the gods. He copulated with his hand and gave birth to Shu and Tefnut (air and moisture), who in turn bore Geb and Nut (earth and sky). Shu then separated Geb and Nut and raised Nut to her present position.[35]

32. Alexander Heidel assigns it this date in *The Babylonian Genesis,* 2nd ed. (Chicago: University of Chicago Press, 1951), p. 14. See also A. E. Speiser's introduction to "The Creation Epic," in *Ancient Near Eastern Texts Relating to the Old Testament,* 3rd ed. by James B. Pritchard, (Princeton: Princeton University Press, 1969), p. 60. More recently Assyriologists have tended toward a later date. W. G. Lambert opines "not earlier than 1100 B.C." ("A New Look at the Babylonian Background of Genesis," *Journal of Theological Studies* 16 [1965]: 291), while others date it to the Kassite period, late fifteenth to mid-twelfth centuries B.C. (see A. K. Grayson, "The Creation Epic: Additions to Tablets v-vii," in *Ancient Near Eastern Texts,* p. 501).

33. A. E. Speiser, "The Creation Epic: Tablet iv:137-140," in *Ancient Near Eastern Texts,* p. 69.

34. Such outlines of Egyptian and Mesopotamian conceptions of the cosmos can suggest that they were ancient counterparts to modern scientific descriptions. That is misleading. Although no doubt based in part on phenomenological observations, they were integral to and can in no way be separated from the religious (mythical) matrix in which they were embedded. Theological and technical cosmic speculations had not yet become differentiated in them.

35. For concise guides through the fluid and (to us moderns) confusing Egyptian

Although other cosmological notions were present among Israel's neighbors, the conception that "heaven and earth" were located between supracelestial and subterranean oceans was widespread among them. It is understandable, then, that the author of Genesis 1 speaks of the waters above the heavens, the starry heavens themselves, the earth (land, surrounded by water), and the waters below the earth (land). There are also similarities between the ways in which these elements are said to have been separated in the Egyptian and Mesopotamian cosmogonies and the corresponding Genesis account, in which a primeval mass of water is divided by the insertion of a "firmament" in which the Sun, Moon, and stars were subsequently set, and the waters below are "gathered to one place" to let the dry land appear. It might be questioned if the "sea" that thus surrounded the "earth" in the Genesis account was viewed as being linked with a subterranean ocean as in Egyptian and Mesopotamian conceptions, since direct reference in Genesis 1 seems to be to the visible seas swarming with life. However, the fact that the biblical account of the Noahic flood speaks of the earth being engulfed by the opening of the floodgates of the heavens and the bursting forth of the springs of the great deep (Gen. 7:11; 8:2) strongly suggests such linkage.[36] Thus the Genesis 1 account of creation contains fundamental features indicating how extensively the author utilized the cosmological concepts of the ancient Near East dating from the third and second millennia B.C. *Even the specific content of God's several creation decrees as here recounted reflects this ancient view of the structure of the cosmos,* a conception of the physical world that differs radically from what is now known of the earth and the heavens.

Polemic against Ancient Near Eastern Mythic Theologies

But while Genesis 1 reflects basic affinities with contemporary notions of cosmic structure, its *theology* involves a break so radical with that of the ancient myths that it is not at all surprising that many in Israel found

myths with cosmogonic implications, see John A. Wilson, "Egypt: The Nature of the Universe," in *Before Philosophy,* by Henri Frankfort et al. (Baltimore: Penguin, 1949), pp. 39-70; and Rudolf Anthes, "Mythology in Ancient Egypt," in *Mythologies of the Ancient World,* ed. S. N. Kramer (Chicago: Quadrangle Books, 1961), pp. 17-92.

36. For other references or allusions to the upper ocean, see Deut. 28:12; 2 Kings 7:2, 19; Job 38:8, 22, 25; Ps. 42:7; 104:3; 148:4; Isa. 24:18; Jer. 10:13 (51:16); Amos 5:8; 9:6; Mal. 3:10. For other references or allusions to the subterranean ocean, see Gen. 49:25; Deut. 33:13; Job 38:16; Ps. 18:16; 24:2; 32:6; 42:7; 69:2, 14; 89:9; 93:3-4; 124:4-5; 136:6; 144:7; Prov. 8:28; S. S. 8:7; Isa. 17:12-13; 27:1; Jer. 51:42; Lam. 3:54; Ezek. 26:19.

it hard to appropriate. Here was a view of God, humanity, and world so alien to that of all other peoples, so thorough and fundamental in the reorientation it demanded, that one needed, as it were, to be born into another world to understand it. No wonder the struggle to liberate Israel from the tyranny of the gods proved so difficult and so protracted. The time came when the idols could be ridiculed as mere blocks of firewood (Isa. 44:14-17), as lies and falsehoods (Ps. 4:2; 40:4; Isa. 28:15; 44:20; Jer. 10:14-15; 13:25; Amos 2:4), as empty nothings (1 Sam. 15:23; Ps. 24:4; 31:6; Isa. 41:29; 66:3; Jer. 18:15) that for all their supposed powers are blind and deaf and dumb and powerless (Ps. 115:5-7; 135:16-17; Isa. 46:1-2, 6-7; Jer. 10:5). Yet there was a reason why the prophets and psalmists heaped such ridicule on the idols, denouncing them as shameful and despicable (Deut. 27:15; 32:16; 1 Kings 11:5, 7; 2 Kings 23:13; Isa. 44:19; Jer. 3:24; 4:1; 7:30; 11:13; Ezek. 5:11; 7:20; Hos. 9:10).[37] For many an Israelite, as for all the peoples about them, the mysterious powers seemingly resident in all things that impinged on humankind and affected their lives were gods that must be honored, placated, and entreated. The first stipulation of the basic document of God's covenant with Israel was no empty rhetoric: "You shall have no other gods besides me" (Exod. 20:3).

The author of Genesis 1 was not merely transmitting a tradition or conveying information or summarizing a common Israelite theology concerning origins on this side of a spiritual struggle long won and laid to rest. His pen served to further that struggle, to break the power of ages-old religious notions that still held many in thrall. He was not grappling with issues arising out of modern scientific attempts to understand the structure, forces, processes, and dimensions (temporal and spatial) of the physical universe. He was not interested in the issues involved in the modern debate over cosmic and biological evolution. His concerns were *exclusively religious.* His intent was to proclaim knowledge of the true God as he has manifested himself in his creative works, to proclaim a right understanding of humankind, world, and history that knowledge of the true God entails—and to proclaim the truth concerning these matters in the face of the false religious notions dominant throughout the world of his day.

In them (i.e., the false religions) the first generation of gods came forth from the primordial watery mass; in his word God is before all, and the primeval waters are his creation. In them the fundamental cosmic entities are gods, and their emergence involved divine procreation; in his word the cosmos is composed of an ordered set of created structures

37. Consider also Elijah's ridicule of Baal on Mt. Carmel (1 Kings 18:27).

fashioned by God's creative word. In them there are gods of the celestial realm, gods of the earth (land), gods of the seas, and gods of the world below; in his word there is one God, who is Creator and Lord of all. In them (especially the Mesopotamian and Canaanite religions) the present ordered world was established only after its creator had conquered the chaos deity or deities, and it remained under the threat of the resurgence of those chaos powers; in his word God sovereignly shaped the plastic medium he had created into the world that now is, and there are no cosmic powers that can threaten to undo it. In them fierce monsters of the deep lurk in the chaos waters (the seas) that surge along the outer bounds of humanity's fragile world; in his word these are mere creatures to be catalogued with the fish of the seas, mere playful denizens of the ocean depths (cf. Ps. 104:26). In them the Sun, Moon, and stars are deities that powerfully affect events on earth (the roots of astrology); in his word the heavenly bodies are but the greater and lesser lights that together with the stars give light on the earth and govern day and night and the seasons—nothing more. In them the mysterious powers of life and its generation involve participation in and manifestation of the divine, since procreation belongs primordially to the gods; in his word creaturely life is not an extension of God's life, and procreation is by God's blessing of living creatures. In them there are certain places in the earth (such as the sites of temples and by extension the cities that maintain them) that are inherently sacred, since they were appointed as such in the beginning and their sacredness belongs to the cosmic order; in his word the whole of creation is the holy kingdom of God. In them humans are but abject slaves and pawns in a metropolis of the gods[38]—except the king, who as a man of power and the representative of the gods participates in the divine;[39] in his word humanity is the crown of creation and all humans alike are, while of earth, fashioned in God's image and appointed to a royal station in the creation.

To be sure, the function of Genesis 1 is not exclusively polemic, and interpreters are not wholly agreed on all the overt indications of polemical intent present in the narrative. But an attentive reading of the whole against the background of the several myths of the ancient Near

38. The ancient myths typically employed the concept of a landed estate or a kingdom, but we use the metaphor of the metropolis to better convey to modern readers the impression left by the religious literature of the ancient Near East that humans occupied a crowded world of the gods, a world in which they played only the role of menial household slaves who were little valued and were lightly trampled on in the push and tug of constant power plays in the highly stratified divine society.

39. Conceptions of the manner of this participation varied.

East discloses a view of God, humanity, and world that, whatever its more or less incidental affinities with conceptions abroad in Israel's environment, stands in striking opposition to almost all that those religions had in common.[40]

Creation as Royal Acts of the Great King

One important feature shared with the other religions of that time and place is the author's use of the political metaphor of divine kingship.[41] In *Enuma elish* Marduk is elevated to kingship among the gods as a result of his victory over Tiamat. It is as king among the gods that he creates the present world order. In Canaanite mythology, El is king over the gods but Baal is elevated to kingship over the world after his defeat of the gods Yamm (Sea) and Mot (Death)—the one a threat to earth (land), the other a threat to life. Egyptian mythology conceived of the Sun god Re (later Amen-Re) as king among the gods (and the king of Egypt as "son of Re"). Similarly, Genesis 1 pervasively evokes the kingship of the Creator.

Here we can do little more than note the several elements of the narrative that reflect this royal metaphor. God's creative words are presented in form and function as royal decrees (cf. Gen. 42:18-20; 2 Kings 12:5; Ezra 1:3-4; 6:3-5; 7:23). His naming of "day," "night," "sky,"

40. For discussion of the polemical elements in Gen. 1, see, in addition to the commentaries, Nahum M. Sarna, *Understanding Genesis* (1966; reprint ed., New York: Schocken Books, 1970), pp. 1-36; and Gerhard F. Hasel, "The Polemical Nature of the Genesis Cosmology," *Evangelical Quarterly* 46 (1974): 81-102.

Regarding the religion of Israel's neighbors, see Frankfort et al., *Before Philosophy;* William F. Albright, *Yahweh and the Gods of Canaan* (Garden City, NY: Doubleday, 1968); E. Theodore Mullen, Jr., *The Divine Council in Canaanite and Early Hebrew Literature* (Chico, CA: Scholars Press, 1980); Conrad E. L'Heureux, *Rank among the Canaanite Gods: El, Baal, and the Repha'im* (Missoula, MT: Scholars Press, 1979); *Mythologies of the Ancient World,* ed. Samuel N. Kramer (Garden City, NY: Doubleday, 1961); Samuel N. Kramer, *From the Poetry of Sumer: Creation, Glorification, Adoration* (Berkeley and Los Angeles: University of California Press, 1979); A. Leo Oppenheim, *Ancient Mesopotamia: Portrait of a Dead Civilization,* 2nd ed. (Chicago: University of Chicago Press, 1968), pp. 171-227; and Thorkild Jacobsen, *Toward the Image of Tammuz and Other Essays on Mesopotamian History and Culture,* ed. William L. Moran (Cambridge: Harvard University Press, 1970), and *The Treasures of Darkness: A History of Mesopotamian Religion* (New Haven: Yale University Press, 1976).

For more general discussions of mythical religious and cosmological conceptions, see Mircea Eliade, *Cosmos and History: The Myth of the Eternal Return* (New York: Harper & Row, 1959), and *The Sacred and the Profane* (New York: Harcourt, Brace & World, 1959).

41. If T. Jacobsen is right, the political metaphor was first applied to the realm of the gods in Mesopotamia in the early part of the third millennium B.C. ("Third Millennium Metaphors: The Gods as Rulers, the Cosmos as Polity," in *The Treasures of Darkness,* pp. 77-143).

"land," and "seas" is a royal expression of lordship over these basic cosmic structures of time and space.[42] And the report of God's assignment of spheres of rule to the Sun and Moon (Gen. 1:16-18) portrays him as sovereign overlord.

At the apex of the account we find a number of especially intriguing evocations of kingly action. First, before proceeding to the creation of humankind, God announces his momentous intention to do so. Both in itself and in the language employed, this announcement recalls the scene in a royal council chamber in which a king announces his impending action to the members of his court (cf. Gen. 3:22; 11:7; 1 Kings 22:19; Isa. 6:8). This passage has occasioned much speculation,[43] but the presence of so many indicators of the royal metaphor surely decides the matter. That humanity is to be created in God's "image" and "likeness" and assigned rule over the creatures in the earth further signifies God's kingly status. "Just as powerful kings [in the ancient Near East], to indicate their claim to dominion, erect an image of themselves in the provinces of their empire where they do not personally appear, so man is placed upon earth in God's image as God's sovereign emblem. He is really only God's representative, summoned to maintain and enforce God's claim to dominion over the earth."[44] Moreover, God's assignment of specific provisions for humanity (vv. 29-30) recalls the royal assignment of food at the king's table (cf. Gen. 43:34; 47:22; 2 Sam. 9:7, 13; 19:28; 1 Kings 2:7; 2 Kings 25:29-30; Jer. 52:33-34; see also Ps. 23:5). Finally, God's blessing of the seventh day (for the sake of man) after finishing his creation of "heaven and earth" has parallels in David's blessing of the people after establishing the ark of God in its tent in Jerusalem

42. "In the ancient Oriental view the act of giving a name meant, above all, the exercise of sovereign right (cf. 2 Kings 23:34; 24:17)," notes von Rad. "Thus the naming of this and all subsequent creative works once more expresses graphically God's claim of lordship over the creatures" (*Genesis*, p. 51).

43. Speculations have mainly centered on whether the text (a) presupposes a plurality in the Godhead (an anticipation of the doctrine of the Trinity), (b) reflects echoes of polytheistic traditions, (c) is an example of the Hebraic plural of majesty, or (d) refers to the angelic host (as in 1 Kings 22:19; Job 1:6, 2:1; Ps. 89:7; Isa. 6:2, 8; and elsewhere). The last is the most likely. To the passages already noted may be added Jer. 23:18, 22, where the true prophet is depicted as one who has been made privy to God's announcements in the heavenly council (cf. Job 15:8) concerning his impending actions on earth. A striking parallel is to be found in Gen. 18: the LORD brings two representatives of his heavenly council with him on his visit to Abram's tent, where he announces the birth of the promised son on the one hand and his imminent judgment on Sodom and Gomorrah on the other. Concerning Gen. 1:26, see M. G. Kline, *Images of the Spirit* (Grand Rapids: Baker Book, 1980), p. 22.

44. Von Rad, *Genesis*, p. 58. See also H. H. Wolff, *Anthropology of the Old Testament* (Philadelphia: Fortress Press, 1974), pp. 160-61.

(2 Sam. 6:18) and in Solomon's blessing of the people after dedicating the temple (1 Kings 8:55).[45]

Genesis 1:1–2:3 tells the story of the first acts of God. It prefaces the biblical account of what God has done and is doing in history in pursuit of his purposes with the creation. No metaphor so permeates that following account as does the political metaphor of king and kingdom. In narrative, prophecy, and psalmody, in Gospels, epistles, and Apocalypse, God's kingship and kingdom are presupposed, proclaimed, invoked, revered, and sung. Human experience provided no analogue—not then, not now—quite so appropriate to convey the nature of the God-world relationship; no other basic metaphor approached it in the scope and degree of its correspondence with the reality it was used to represent and thus in the wealth of creative ways it could be exploited for theological and religious expression. Not surprisingly, then, God's first acts also, his sovereign creating acts, are depicted as the initial edicts of the Great King by which he founded and ordered his kingdom.

The contextual appropriateness of such a presentation of creation is even more evident when note is taken of the fact that this account serves immediately as the prologue to an extended narrative block (the Pentateuch) that is dominated and thematically unifed by a series of divine covenants (with Noah, with the patriarchs, with Israel) all of which are modeled after ancient Near Eastern royal administrative (legal) instruments—all of which present God as the Great King administering his kingdom.[46]

In Genesis 1:1–2:3 the several evocations of the basic political

45. Cf. also Moses' blessing of the people after the completion of the tabernacle (Exod. 39:43).

The parallel between God's blessing the seventh day and these royal blessings appears to be warranted by the fact that the visible creation is elsewhere viewed as God's glory-filled temple (Ps. 29:9; cf. Ps. 104:1-3; Num. 14:21; Ps. 72:19; Isa. 66:1-2a), and the tabernacle and temple were viewed as microcosms of the visible universe—as was true of temples throughout the ancient Near East. See M. G. Kline, *Images of the Spirit,* pp. 35-42. Here we may add that if Kline's intriguing proposals are correct, both the hovering Spirit of v. 2 and the seventh-day rest of 2:2-3 are additional exponents of the Creator's kingly status, the former as an emblem of the theophanic glory cloud that manifested God's "glory of royal majesty" and enclosed his throne (cf. Ezek. 1:26-28), and the latter as "a royal resting, an enthronement on the judgment seat" (*Images of the Spirit,* pp. 13-20, 111).

46. These covenants are of two distinct kinds: the royal land-grant type and the Suzerain-vassel type. On the covenants, see G. E. Mendenhall, *Law and Covenant in Israel and the Ancient Near East* (Pittsburgh: Biblical Colloquium, 1955); K. Baltzer, *The Covenant Formulary in Old Testament, Jewish, and Early Christian Writings* (Philadelphia: Fortress Press, 1971); D. J. McCarthy, *Treaty and Covenant* (Rome: Pontifical Biblical Institute, 1963); M. G. Kline, *Treaty of the Great King* (Grand Rapids: Eerdmans, 1963), and *By Oath Consigned* (Grand Rapids: Eerdmans, 1968); and M. Weinfeld, "The Covenant of Grant in the Old Testament and in the Ancient Near East," *Journal of the American Oriental Society* 90 (1970): 184-213.

metaphor of kingship are subtle and allusive. The total effect, however, is to depict the absolute kingship of the God of Israel over all realms visible and invisible and over all cosmic regions (the heavens, the waters above the heavens, the earth, and the seas below) and their denizens. In fact, this implicit claim constitutes one of the most powerful affirmations of the first chapter of Scripture.

Later in this chapter we will explore more fully some implications of this royal metaphor that are especially relevant to the concerns of this volume; here we only note it as a point of contact with the cosmogonic myths of Israel's environment and a theologically significant element in the manner in which Genesis 1 proclaims God's creative acts.

Narrative Perspective

Consonant with the use of the royal metaphor is the narrative perspective the author has chosen. By perspective we mean the angle from which the narrator views the events recounted, the location in time and place he assumes relative to the events of his "story." Commonly a narrator places himself on the scene of these events. For example, when the prophet Micaiah reports to Ahab that he "saw [presumably in a vision] all Israel scattered on the hills like sheep without a shepherd" (1 Kings 22:17), his perspective is that of one standing on the hills of Israel where the "sheep" were scattered. But when he tells of his vision of God's action in the divine council (vv. 19-22), he stands as it were in God's heavenly council chamber and witnesses what transpires there. Similarly, the author of Job initially takes his stand in God's heavenly court (1:6-12; 2:1-6) and then with Job on earth (1:13-21; 2:7-10). Throughout the following dialogues he remains near that sufferer, even when God discloses himself to Job out of the storm (chaps. 38–40).

In Genesis 1:1–2:3, the author recounts the creation of heaven and earth as if he had stood in the very presence of the Creator as he issued his decrees, assessed his works, announced his intentions concerning the creation of humankind, pronounced his benedictions, assigned provisions for his creatures, and consecrated and blessed the seventh day. He "reports" as one who had been an eyewitness to events in God's executive chamber. By contrast, in Genesis 2:4ff. the narrator stands with his feet on the earth, in the arena of human history. To be sure, the author of 1:1–2:3 had not been caught up into the heavens (like the prophets—see 1 Kings 22:19ff.; Jer. 23:16ff.; cf. 2 Cor. 12:2; Rev. 4:1-2) to witness events in the heavenly realm—the Creator does not effect his creation work from afar, as witness the Spirit "hovering over the surface of the waters" (Gen. 1:2). Nevertheless, the author writes as one made privy to the inner precincts of God's executive actions.

There was, of course, "in the beginning" no other place to stand, no other vantage point. But the narrator's perspective does not shift until 2:4. Still, so near at hand is the Creator that once he has created light and named "day" and "night," the distribution of his creative acts (as the narrator recounts them) is governed by that sequence. And since the narrator knows nothing of the spherical shape of planet Earth, its daily rotation, its annual revolution around the Sun, and that star's situation in the Milky Way Galaxy, only one of many billions of galaxies—and it is the narrator's perspective that we are dealing with here[47]—the day-night alternation has for him absolute cosmic status. Nevertheless, he writes as one who has witnessed the creation from the arena of God's creative activity, not as one whose feet are planted on "earth" (land).

The Seven Days

Although this account is narrated in the form of a series of distinct divine acts distributed over a period of seven "days" and in a manner that approximates the chronicling of royal acts, it ought not to be read as though it were a record of events in the court of an earthly king. What occurs in the arena of God's action can be storied after the manner of human events, but accounts of "events" in that arena are fundamentally different in kind from all forms of historiography. As representations of what has transpired in the divine arena, they are of the nature of metaphorical narrations. They relate what has taken place behind the veil, but translate it into images we can grasp—as do the biblical visions of the heavenly court. However realistic they seem, an essential "as if" quality pervades them. Still, because what is done in the divine arena can be storied, the story of what happens there can be linked in narrative continuity with accounts of events in earthly history, as is done, for example, in Genesis 11:1-9; 1 Kings 22; Job 1–2; and Isaiah 6.[48] Such

47. To assume, as some have done, that God himself is the narrator is to ignore the fact that the author makes no claim that Gen. 1 is a direct oracle from God (after the manner of prophetic oracles). He presents it, like the rest of Genesis, as his own composition. Divine inspiration of Gen. 1 is not to be thought of as divine dictation—no more for it than for what follows in Genesis, or for the Gospel of Matthew, or for one of Paul's epistles.

48. Such linkage occurs also in the parable of the rich man and Lazarus (Luke 16:19-31).

continuity also links the stories of events in these two arenas in Genesis 1 and following.

Both the manner and form of the proclamation of creation in Genesis 1:1–2:3 belong, therefore, to the metaphorical character of the presentation. The author stories God's creation of the world "in the beginning" as truly a series of divine *acts*—but as a series of truly *divine* acts performed not in human history but in the divine arena. He stories "events" that are in themselves inaccessible to humans, inaccessible not only as information (since no human witness was there) but *conceptually* inaccessible. Through the realism of his account he makes humanly conceivable what in itself is beyond human apprehension.

Hence the "days" of Genesis 1:1–2:3 are not presented as the first seven days of the story told in Genesis 2:4ff. In the terms of this account they mark "time" in the arena of God's creation activity. That was for the author a special "time," a "temporal" sequence belonging uniquely to the arena of the creation acts of God as storied. From the perspective of this account, these seven "days" are a completed time—the seventh day does not give way to an eighth. It was completed time because in it the unique works of creation were finished and the Creator "rested" from creating. And because this narrative stories unique events in a unique arena and a unique "time," the lack of correlation between the chronological sequences of 1:1–2:3 and 2:4ff. involves no tension.

The ages-old debate concerning the length of these creation "days" (whether they were twenty-four hours or some indefinitely long period of time, whether or not they were all of equal length) has continued unabated to the present. And as long as it is supposed that somehow these mark time periods (whether long or short) directly continuous with the days and years of our conventional calendars, the issue cannot be resolved. Surely there is no sign or hint within the narrative itself that the author thought his "days" to be irregular designations—first a series of undefined periods, then a series of solar days—or that the "days" he bounded with "evening and morning" could possibly be understood as long eons of time. His language is plain and simple, and he speaks in plain and simple terms of one of the most common elements in humanity's experience of the world. The speculations that have continued to fund the endless and fruitless debate have all been triggered by concerns brought by interpreters to the text, concerns completely alien to it. In his storying of God's creative acts, the author was "moved" to sequence them after the manner of human acts and "time" them after the pattern of created time in humanity's arena of experience.[49] Such sequencing and dating belonged integrally to the whole fabric of his account (the heavenly King commanding his realm into existence and ordering its internal affairs) whereby he made imaginable the unimaginable.

Recognition that Genesis 1:1–2:3 presents a storied rather than a historiographical account of creation reinforces the conviction of many interpreters that the topical selection and arrangement, as well as the sabbatical distribution of the acts of creation, are governed by the demands and logic of the purpose of the presentation in the historical context of the author and the literary context of Genesis. The account's geo-anthropocentic focus reflects the geocentric world picture of the times and comports with the anthropocentric concerns of the Creator that dominate the narrative that follows. As noted above, the creation of light and the division of light and darkness into the alternation of day and night accords with the ancient perception of that alternation as the basic structure of cosmic time. God's division of a primeval mass of water into the ocean above and the ocean below and the view that the "firmament" is in some sense a cosmic structure holding back the heavenly ocean while the earth is a cosmic region that emerges from the waters below—these are reflexes of the ancient Near East world picture rather than of reality as we now know it. That the account is topically progressive (given the ancient world picture) and climactic, with humanity portrayed as the crown of creation, has often been noted, as have its structural symmetry and parallelism.[50]

If we ask why the narrator distributed the creation acts of God

49. Accordingly, one might as well attempt to correlate the "days" of Job 1:6 and 2:1 with some particular days in Job's life as to seek to correlate the "days" of Gen. 1 with those of our calendar (whether we believe them to be "days" of twenty-four hours or of a hundred millennia).

50. The symmetry and parallelism can be briefly outlined as follows:

Day 1 Creation of light and separation of day and night.

Day 2 Creation of the "firmament" and separation of the waters above from the waters below.

Day 3 Creation of the "earth" (dry land) by "gathering into one place" the waters below; calling forth plant life from the earth.

Day 4 Creation of luminaries to govern day and night.

Day 5 Creation of life forms to fill the waters below and to fly above the earth across the face of the "firmament."

Day 6 Calling forth land animals from the ground; creation of man to rule over all forms of life.

Day 7 Divine rest.

Seven divine assessments that the creation is "good" (1:4, 10, 12, 18, 21, 25, 31).

over a period of time and why he then employed a seven-day (sabbatical) structure, it must be admitted that we are not expressly told.[51] But our conclusions in this regard need not be arbitrary. Concerning the first, we may note that all the Near East theogonic and cosmogonic myths speak of a series of events. This may well reflect the limits of human conceptuality—namely, that the origins of a reality so complex as the world must have required a series of divine acts. Moreover, a creation story that only accounted for a divinely ordered structure of the cosmos and not also for divinely appointed temporal orders would seem singularly incomplete, especially a creation account that prefaces a recounting of the history of the divine-human relationship in which the destiny of the created world is decisively affected.

As regards the seven-day structure, any other temporal order would appear to have been unfitting in that ancient world. Throughout the ancient Near East the number seven had long served as the primary numerical symbol of fullness / completeness / perfection, and the seven-day cycle was an old and well-established convention. Out of the many examples available we may note from Ugaritic literature (1) the seven-day journey of King Keret to the city of Udum the Great and his seven-day seige of that city,[52] (2) the seven days required for completing Baal's royal palace,[53] and (3) King Danel's seven-day appeal to the gods and his subsequent seven days of feasting.[54] According to the Mesopotamian Epic of Gilgamesh, it took seven days to build Utnapishtim's "ark," and the flood that followed raged for seven days and took a similar period of time to subside.[55] Functionally, the sabbatic structure of creation "time" added symbolic reinforcement of the explicit themes of the completeness of God's creative work and the "goodness" of the created realm. It also introduced a structure of time that is not governed by the Sun and Moon (humans have little choice but to honor solar and

51. St. Augustine is only the most illustrious theologian to affirm for various reasons (his were philosophical) that in actuality creation must have been instantaneous.

52. "The Legend of King Keret," *Ancient Near Eastern Texts*, pp. 144-45.

53. "Poems about Ba'al and Anath," *Ancient Near Eastern Texts*, p. 134.

54. "The Tale of Aqhat," *Ancient Near Eastern Texts*, p. 150.

55. "The Epic of Gilgamesh," *Ancient Near Eastern Texts*, p. 94. From the Old Testament come such examples as the seven days Noah waited in the ark for the flood to begin (Gen. 7:10), the two seven-day periods he waited between his sending forth of doves after the ark came to rest (Gen. 8:10, 12), Moses' seven-day wait on Mt. Sinai for the LORD to come (Exod. 24:16), the seven-day siege of Jericho (Josh. 6), the seven days in which the armies of Ahab and the Arameans faced each other before joining battle (1 Kings 20:29), and the seven-day wedding feast (Judg. 14:12, 17; cf. Gen. 29:27). In addition, the seven-day feasts of Passover-Unleavened Bread and Tabernacles are well known.

lunar time periods—cf. Ps. 104:23; 127:2) but is grounded solely in the divine paradigm of the Creator at work. Humans honor this sabbatic pattern in their stewardship of the creation only as they acknowledge the Creator as their Lord.

Both ancient and modern attempts to integrate the "days" of the creation week with time as measured in our ordinary human experience—whether as the first seven days of our calendar or as eons corresponding to geological ages—have been based on a problematic assumption concerning the biblical text. The presupposition that the events depicted in Genesis 1 can be placed on the human calendar fails, we believe, to take properly into account both the narrative perspective and the theological focus of the creation story. When these are taken into account, we gain a basis for understanding why the many concordist theories proposed by Christians to harmonize Genesis and geology have proved to be so unfruitful. For twentieth-century Christians to expend considerable effort to revive these theories or to give them the appearance of scientific merit strikes us as a step backward, a step in the direction of associating Christian faith with theories that long ago failed the test of careful scrutiny.

Literary Genre

Our discussion of Genesis 1:1–2:3 raises the question of the literary genre (type) the passage represents. For many readers of the Bible this question never arises. Because on its surface it seems to tell of a series of divine acts undertaken in a numbered sequence of "days," because it introduces an account of the history of the God-humanity relationship, and perhaps because of notions about the "simplicity" of ancient modes of thought, some readers assume that here, too, they are reading a narrative that belongs in some sense to the genre historiography or unadorned chronicle. However, as we have argued, such a reading cannot be sustained in light of what we know about the whole of Scripture.

But if this opening word of Scripture is not simply a chronicle of historical particulars, to what genre does it belong? There has been no lack of proposals. To classify it as myth is to employ a term of such a varied usage that it is virtually meaningless—or worse, misleading, since it obscures the author's sharp opposition to the basic conceptions of God, humanity, and world embodied in the myths current in his day. To some the passage suggests a creation hymn, but it clearly does not employ the same poetic mode, linguistic conventions, and doxological tone as other such hymns in the Old Testament (cf. Ps. 8; 19:1-6; 29; 65:9-13; 104; Amos 4:13; 5:8-9; 9:5-6). The suggestion that it was originally a cult liturgy composed for a New Year festival celebrating creation equally lacks cogency.[56] Not only would such a liturgy most likely be more hymnic than the composition we have, but there is no compelling evidence that Israel had the sort of New Year festival that the proposal requires. Gerhard von Rad is nearer the truth when he insists that "Whoever expounds Gen., ch. 1, must understand one thing: this chapter is . . . doctrine."[57] It is, to be sure, doctrine in a narrative mode, but that is both consistent with all accounts of creation in the ancient world[58] and appropriate to the function of Genesis 1 as prologue to a narrative of God's engagements in human history. Still, even as narrative it is unique in that, while it moves toward climax, both conflict and tension are notably absent. In short, its literary type, as far as present knowledge goes, is without strict parallel; it is *sui generis.*[59] As an account of creation it supplies for the Pentateuch what for the religions of Israel's neighbors were supplied by their mythic theogonies and comogonies. Consistent with its theme, however, its form (an unadorned "objective" account of "dated" events, using formulaic language) suggests that of a "record," recounting what transpired in God's royal council chambers. It reads like a daybook kept by a recorder of royal executive actions.[60]

However unique its genre and whatever its date relative to the rest of Genesis, it was undoubtedly composed to serve as prologue to 2:4ff. It provides for that account of the historical unfolding of the

56. For an example of this argument, see S. H. Hooke, *Middle Eastern Mythology* (Baltimore: Penguin Books, 1963) pp. 119-21.

57. Von Rad, *Genesis,* p. 45. In accordance with the critical consensus that Gen. 1:1–2:4a belongs to the latest revision of the Pentateuch carried out by a priestly school of theologians in the exile and early postexilic period, von Rad speaks of it as "Priestly" doctrine.

58. All the myths of creation which may have been known by the author were, in fact, "doctrine" in the form of stories.

59. Robert Alter's statement concerning biblical narrative in general (following Herbert Schneidau, *Sacred Discontent* [Baton Rouge, 1977])—namely, that it appears to be a "new medium which the ancient Hebrew writers fashioned for their monotheistic purposes"—seems surely to be true of Gen. 1. See *The Art of Biblical Narrative* (New York: Basic Books, 1981), p. 25.

60. Although no such daybook has been recovered from the ancient Near East, it is known that royal records of various kinds were kept: note the biblical references to royal "recorders" (2 Sam. 8:16; 2 Kings 18:18; 2 Chr. 34:8; etc.), to "the book of the annals of the kings of Israel" (1 Kings 14:19; etc.), to "the book of the annals of the kings of Judah" (1 Kings 14:29; etc.), and to "the book of the annals" of the Persian Kings (Est. 2:23; 10:2; cf. 6:1).

God-humanity relationship the ultimate horizon and the fundamental doctrine concerning God, humanity, and world: of God, that he is one, personal, without beginning, the sovereign Creator of all things visible and invisible, the absolute, wise, and benevolent Lord over all; of the world, that it is creature, ordered, good, purposive, dependent, designed as a kingdom of life, and subject to God's sovereign rule; of humanity, that it is creaturely, of the earth yet created in the image of God, God's steward of the visible creation, the special object of God's care, and the epicenter of the divine enterprise. Only of this God, this world, and this humanity could the following story be told.

To read Genesis 1:1–2:3 as a piece of divinely revealed "historiography" disclosed to humanity's first pair and transmitted by tradition to the author of Genesis will no longer do. To do so is to suppose an unbroken transmission of tradition that can no longer be assumed. It is also to ignore this account's many affinities with third to first millennium Near Eastern notions about the structure of the cosmos, its massive polemical thrust against the mythic theologies of the day, its reflection of royal ideology (which can hardly predate the third millennium B.C.), its narrative perspective, and its distinctive literary form. While Genesis 2:4ff. presents an account of God's ways with humankind in the arena of human history, the grand overture that preceeds it presents not historical or scientific data but the fundamental theological (and its related anthropological and cosmological) context of that drama. It is the story behind *the* biblical story. And more than that, it is the story behind all cosmic, terrestrial, and human history. Only when we hear it so do we catch its theme. Only when we hear it so can we rightly understand the long story that follows—and the story that unfolds as humanity investigates the creation.

THE CREATION DECREES AND PROVIDENCE

The Theological Distinction between Creation and Providence

Systematic theologians have long taught us to make a sharp distinction between creation and providence. Louis Berkhof states the common view succinctly:

> Creation is the calling into existence of that which did not exist before, while providence continues and causes to continue what has already been called into existence. In the former there can be no cooperation of

> the creature with the Creator, but in the latter there is a concurrence of the first Cause with second causes.[61]

With many others, the noted Reformed theologian Herman Bavinck argued that Christian theology must maintain this distinction between creation (the divine bringing into existence) and providence (the divine conservation and government of the creation) over against all forms of pantheism on the one hand (which fail to distinguish God and world) and all forms of deism on the other (which posit an independent existence of the created world). But he also contended that theology must keep these conceptually distinguished in order to be true to Scripture. He held that references to God's resting (Gen. 2:2; Exod. 20:11; 31:17) serve to indicate God's transition from creating to preserving and governing—that is, a transition from the time of God's bringing new realities into being to the time of his preserving and governing what he had created. And he insisted on this distinction even though he had to acknowledge that creation and providence are inseparably related, that they are, in fact, so closely bound together that Scripture even uses creation language to speak of providence (he refers to Ps. 104:30; 148:7; Isa. 45:7; Amos 4:13).[62]

Abraham Kuyper, Bavinck's even more widely known contemporary, took a somewhat different tack. In his effort to find language appropriate for and adequate to the elucidation of divine providence while maintaining a distinction between the origination and the continuation of the created, he concluded that language fashioned for talk of human activity provided no serviceable analogues for divine preservation. Hence he settled on a theological phrase that already had a long history—*creatio continuata.*[63] By means of this terminology he sought to safeguard theologically the inseparable link between origination and preservation as divine works and to maintain the total dependence of the created on the Creator. "We must definitely insist that providence is a *creatio continuata,* to be understood in the sense that from the hour of original creation until now, God, the Lord, has done the same thing as in the moment of creation: he has given all things power of existence through his power."[64] He had to hedge in this language carefully,

61. Berkhof, *Reformed Dogmatics* (Grand Rapids: Eerdmans, 1941), p. 167.

62. Bavinck, *Gereformeerde Dogmatiek,* Tweede Deel (Kampen: J. H. Kok, 1918), p. 136.

63. Kuyper, *Dictaten Dogmatiek,* "Locus de providentia," "Relatio quae Providentia cum Creatio intercedit."

64. Kuyper, quoted by G. C. Berkouwer in *The Providence of God* (Grand Rapids: Eerdmans, 1952) p. 62.

however, to exclude the notion of an endless series of *de novo* creations with only the appearance of continuity.

Kuyper's attempt to revalidate *creatio continuata* as a helpful theological concept has found little acceptance, however, and mainly for such reasons as those advanced earlier by Charles Hodge. Hodge insisted that "In the Bible the two things [creation and providence] are never confounded." "It is true," he added, "that the preservation of the world is as much due to the immediate power of God as its creation, but this does not prove that preservation is creation. Creation is the production of something out of nothing. Preservation is the upholding in existence what already is."[65]

Standing as they did within a long theological tradition that had shaped its problematics and concepts in dialogue with many philosophies and heresies, these theologians seem unaware that their sharp distinction between creation and preservation, however conceptually neat and theologically useful, might be a distinction they were sometimes reading back into the Bible. It can hardly be doubted that the distinction itself is present in Scripture (cf. Heb. 1:2-3), but it is quite another matter to claim that it is always and everywhere maintained in the biblical literature. Scripture often speaks of preservation without speaking of creation, but it may well be asked if, when speaking of "creating," Scripture does not at times also include what many theologians have subsumed under preservation—specifically, that God's "creating" act includes his sovereign preservation of the dependent creature.

If B. B. Warfield is right, Calvin's views on these matters were not as rigid as those of most of his theological heirs. According to Warfield, Calvin limited "creation" proper to God's bringing into being *ex nihilo* two things: the original world-stuff and individual human souls. For the rest, God sovereignly formed the several creatures out of the original world-stuff. While he could not quote Calvin directly to this effect, Warfield suggests that Calvin came close to ascribing this secondary forming of creatures to providence—in view of Calvin's "high" view of providence and in view of the fact that he acknowledged the role of secondary causes in the process. In fact, Warfield makes bold to say,

> It should scarcely be passed without remark that Calvin's doctrine of creation is, if we have understood it aright, for all except the souls of man, an evolutionary one. The "indigested mass," including the "promise and

65. Hodge, *Systematic Theology,* vol. 1 (New York: Charles Scribner's Sons, 1883), pp. 578-79.

> potency" of all that was yet to be, was called into being by the simple *fiat* of God. But all that has come into being since—except the souls of men alone—has arisen as a modification of this original world-stuff by means of the interaction of its intrinsic forces. Not of these forces apart from God, of course. . . . And this, we say, is a very pure evolutionary scheme. . . . Calvin doubtless had no theory whatever of evolution; but he teaches a doctrine of evolution.[66]

Warfield's ascription to Calvin of a kind of theistic evolution is probably misleading in the post-Darwin era. But the main thrust of his summary of Calvin suggests that the Reformer's distinction between creation and providence was significantly less sharp than that of the later theologians noted above.

Returning to more recent times, however, we find G. C. Berkouwer still basically agreeing with the consensus quoted above from Berkhof. "The Scriptures," he writes, "compel us to make the distinction between creation and sustenance" (the latter an aspect of providence).[67] And other recent Reformed theologians, such as Emil Brunner[68] and Otto Weber[69] are of the same mind.

66. Warfield, "Calvin's Doctrine of Creation," in *Calvin and Calvinism* (New York: Oxford University, 1931), p. 304-5. See also p. 306 n. 45: "Calvin accordingly very naturally thought along the lines of theistic evolution."

67. Berkhouwer, *The Providence of God*, p. 66.

68. Writes Brunner,

> . . . any one who does not admit the distinction between the creation and the preservation of the created world does not take the fact of creation seriously. The relation of God to that which He has created is not the same as His relation to that which is yet to be created. That which has been created stands actually "over against" God. Henceforth, through the action of God it has an independent existence, even though this independence be a limited one. It depends on a divine thread of preservation above the abyss of nothingness; at any moment God can let it fall into nothingness. But to preserve that which has been created does not mean continually to create it anew; to claim this would mean that it has an actual existence for God, that it has an existence of its own. . . . Now the recognition of a divine preservation of the world as distinct from His creation, does not exclude the truth that God is still actively and creatively at work in a world which He has already created, and which He preserves. (*The Christian Doctrine of Creation and Redemption* [Philadelphia: Westminster, 1952], pp. 33-34)

69. Writes Weber,

> Creation and conservation, or creation and providence, cannot be identified as one, although certainly the Creator is the Conservor and Ruler. The most illuminating reason for this is the simple fact that God's conserving and ruling activity, his providence, by no means takes place "out of nothingness," but conserves created existence as something already extant and active and thus presupposes it. . . . Regardless of the way one proceeds terminologically, the distinction between creation and providence should be quite clearly obvious. Christian theology

If one scans the history of the discussion, it is obvious that theologians have generally held fast to a theological distinction between creation and providence. Simple logic seemed to demand it, and it has served effectively to safeguard the uniqueness of the biblical ontology (by excluding both pantheism and deism) and to preserve God from the charge of being the author of evil.

Does the Bible Keep Creation and Providence Distinct?

However necessary and useful this theological distinction may be, its easy imposition on biblical texts surely needs reassessment.

To begin with, as we have seen, Hodge's insistence that "In Scripture the two [creation and providence] are always kept distinct" clearly needs qualification in light of Psalm 104:30 (God's "creation" of each new generation of living things), Psalm 102:18 (the "creation" of each new generation of worshipers), Ecclesiastes 12:1 (God's "creation" of each individual; cf. Job 10:8-12; 31:15; 33:4; 40:15; Ps. 139:13-15; Prov. 22:2; Isa. 43:7; Mal. 2:10), Isaiah 43:1 (God's "creation" of Israel; cf. v. 15; 27:11; 44:2, 24), and Isaiah 54:16 (God's "creation" of the smith and the destroyer).

But more to the point, God's "Let there be . . ." in Genesis 1 is too narrowly conceived if it be supposed that it stands only as a power word to effect origination, the means by which God brought into being creatures that had then to be maintained by a *new* divine act, another decree from the mouth of God, such as, "Let the created be preserved." There is no such additional word—and that is not because the narrator omitted it but because God's "Let there be . . ." was a sovereign establishing and sustaining word. His creation decree was and is the fundamental preserving and governing word.

Consider, for example, "Let there be light." In the narrative of Genesis 1, this initial decree signalizes a mighty invasion into the world not yet differentiated (of which darkness is the most elemental exponent) by God's creative word, and in terms of the account it makes it

---

in general has never fallen prey to the temptation of assuming the Creator's omnicausality so that the reality of the creature's activity was severely restricted. Conversely, it has never assigned such importance to the creature's own causality that the effectiveness and reality of the Creator were made into nothing more than the initiative of the past. (*Foundations of Dogmatics,* vol. 1 [Grand Rapids: Eerdmans, 1981], p. 305)

possible that all the subsequent creative acts of God be carried out in the full light of day. God's creative works are not deeds of darkness.[70] But the light that is created on the first creation day is not light that emanates from light-giving bodies; these are said to be created later. It is neither a material entity with its own relatively self-sustaining economy, nor is it the product of such an entity. It is a created reality that is originated by and remains totally dependent on the creative "Let there be." The creation decree originates, preserves, and governs light. There is now, today, light in the world, and the alternation of day and night, our author says, not because in the beginning God created a very durable light but because in the beginning *God issued a decree enduring in its effect.*

But the same is true of "let there be a firmament in the midst of the waters, and let it separate the waters from the waters." With that decree the waters above were separated from the waters below, but what was created thereby was no colossal celestial dam fabricated of stainless steel (like the Gateway Arch of St. Louis) to withstand the corrosive effects of the ages, a structure that once fashioned has permanence from its own strength and durability. By God's decree the "firmament" both came to be *and is preserved.*

And no less must be said of "Let the waters under the heavens be gathered together into one place, and let dry land appear." By this decree the seas are even now kept in place (cf. Job. 38:8-11; Ps. 33:7; 104:9; Prov. 8:29; Jer. 5:22).[71]

For various reasons Christian theologians have been so preoccupied with creation *ex nihilo* and so concerned to keep creation and providence distinct that the creation decrees of Genesis 1 have been viewed almost exclusively as originating decrees—a view that has encouraged deistic notions of God's creative activity. But it would seem more faithful to the true function of these decrees to conceive of them as analogous to King Omri's (presumed) decree, "Let there be a royal city

70. See Job 24:13-17; Ps. 82:5; Prov. 4:18-19; Eccl. 2:14; etc. And note Westermann's observation that "The creation of light is put before [the basic cosmic divisions effected in the first three "days"] because it renders possible the temporal succession into which . . . the world is set" (*Genesis 1–11*, p. 112).

71. Perhaps it is from this perspective also that we best understand the participles of the hymnic celebrations of creation: Ps. 104; 147; 148; Amos 4:13; 5:8; 9:6. Note should especially be taken of Ps. 104:5-6:

> For he commanded and they were created.
> He set them in place for ever and ever;
> he gave a decree that will never pass away.

on this hill"—to which we may add, by slight paraphrase of 1 Kings 16:24, "and he built the city and named it Samaria."[72]

The actual initial construction of Samaria's walls, houses, and palaces by Omri's craftsmen would and could effect only the erection of structures that would immediately begin to deteriorate and eventually crumble. But Omri's royal decree, his "Let there be," did more than activate his craftsmen to erect some structures; it founded a royal city. It was not just a work order; it was a decree that a city come into being and endure as a royal city sustained by the king's authority and resources and at the same time by the continuing service of his subjects. The origination and preservation of Samaria as the king's royal city flow from the one decree. Similarly, God's creating decrees originate, preserve, and govern what he created.

And since God's creating word called into being not merely an aggregate of entities but components designed to function within the economy of an integrated realm, the preserving and governing power of the creation word cannot be isolated from the functioning economy of the creatures, from their "concurrence." The creation's economy continues to be governed by the Creator's decrees—his *creation* decrees. God's "decrees" for the rain (Job 28:26), for the celestial realm (Job 38:33; Ps. 148:6; Jer. 31:35-36), and for the sea (Prov. 8:29; cf. Job 38:11; Ps. 104:9) are all *creation* decrees. There is indeed no continuous creating (no *creatio continuata* in its classic sense); there is rather the continuing effectiveness of the Creator's "Let there be . . ." and his "Let the land produce vegetation" and his "Let the waters teem with living creatures" and his "Let the earth produce living creatures."[73]

Significantly, in the case of fish, birds (presumably also land animals), and humankind, God's creating word is immediately accompanied by his blessing word, "Be fruitful and multiply and fill . . ." The self-generating creatures are preserved according to their nature by the continuing effectiveness of the once-for-all divine word of blessing. God's sovereign "Let there be . . ." and his sovereign "Be fruitful . . ." *together* constitute his creation-ordering word for biological life, a word

72. Cf. Cyrus's royal edict concerning the temple at Jerusalem (Ezra 1:2-4).

73. Here it must be observed that those who hold to some evolutionary theory cannot appeal to these words to suggest that they anticipate or hint at organic evolution. These forms of the creation word for living things are rather a reflex of the common observation that plants and trees spring from the ground and that since at death animals and man return to the dust, they must have come from the dust (cf. Gen. 2:7, 19; Job. 4:19; 34:15; Ps. 103:14; 146:4; Eccl. 3:20; 12:7—and notice the association of the womb with "the depths of the earth," Ps. 139:13, 15; cf. Job 1:21; 10:9-11, 18).

that originates, preserves, and governs. The effectuation of this word of blessing can even be called "creation," as when God is said to "create" new generations of living things (Ps. 104:30; 102:18; 139:13; etc.).

The problem that has so long exercised theologians is transcended, it seems, by the more concrete language, the narrative mode, of Genesis 1. There God's royal "Let there be . . ." is portrayed as effective both to originate and to maintain and govern the creation in all its complex yet harmonious economy.

In recounting God's creation decrees, the author of Genesis 1:1–2:3 was not merely telling a story of how a world that once was, "in the beginning," had come to be. His is rather a word about God's issuing a series of creation decrees "in the beginning" that in their continuing sovereign effectiveness account for the world as he and his readers knew it—a world in which light breaks into darkness and overcomes it, bounds it, and even domesticates it; in which the restless seas, however powerfully they may surge, are kept firmly at bay; in which what is above does not threaten to collapse and mingle again with what is below; in which the visible heavens are populated with luminaries that regulate (only) the days and seasons; in which lifeless water and soil nurture a plethora of life; in which humankind, made of dust but endowed with powers and prerogatives unique among the creatures, participates (as steward) in God's rule over his creation—in short, a world that every day, in all its order, vitality, and progressive unfolding, incarnates God's creation words issued "in the beginning."

That the Creator's creation decrees had been at work for billions of years (as humans count time) before the physical universe had attained the form he perceived and become the arena of human history as he knew it was not known by (or made known to) the writer of this creation narrative. Nor was that his concern (or the concern of the Spirit that "moved" him). Cosmological and geological "history" in the modern scientific sense were not in his purview and were not as such at stake in the great religious issues that engaged him. What did concern him was the profusion of gods "feared" by the nations and the need to establish in Israel the uncompromised faith that the God who had called Abraham and rescued Israel from Egypt and bound them to himself in covenant was the only God, the Creator and Lord of all, whose dealings with the fathers and with Israel were pursuant of the divine purposes embodied already in the creation of the world—that all reality stands under the rule of the one Lord and all history is directed toward one goal, the full realization of the kingdom of God.

To limit Genesis 1:1–2:3 and specifically the creation decrees there recounted to a word merely about the once-upon-a-time origin of the visible and invisible creation is in large measure to trivialize it.

THE CREATION AS GOD'S KINGDOM AND MAN AS GOD'S ROYAL STEWARD

Within the conceptual system of the Old Testament the most common metaphor for the created realm is that of the kingdom of God.[74] As has already been observed, the royal metaphor also pervades the Genesis 1 account of creation, evoking the metaphor of kingdom for the realm created. That the visible creation is here viewed as a single ordered and integrated realm is made emphatically clear by the manner in which a progressive sequence of God's creative acts is related. Out of a single primordial mass God calls forth and fashions the visible universe with all its components, relationships, and functions. And as is true of a political kingdom, the creation kingdom is knit together and integrated by its own economy under the rule of God.[75]

Within the economy of that kingdom, humanity was created for and appointed to the office of royal steward under the Creator-King. Having decreed all else into existence, God said, "and now let us make humankind, in our image, according to our likeness, and let them rule[76]

74. In distinction, for example, from that of an organism (Greek cosmology) or a machine (Newtonian cosmology). The data is extensive and varied:

a. God is said to be or is depicted as King over all the earth: Ps. 29; 47; 93; 94; 95; 96; 97; 98; 99; 104; 145; Isa. 6; Jer. 10:10; Dan. 4:37; cf. Exod. 17:16; 1 Chr. 29:12; Ps. 11:4; 22:28; 66:7; 89:1-18; 103:19; 123:1; 146:10.

b. The created realm is occasionally called the "kingdom" of God: 1 Chr. 29:11; cf. Ps. 103:19, 22; 145:11-13.

c. God governs the created realm by his decrees: Job 28:26; Ps. 148:6; Jer. 31:35-36; and his commands: Job 36:32; 37:12; 38:11-12, cf. v. 34; 39:27; Isa. 5:6.

75. For example, Sun and Moon give light in the earth and govern day and night and the seasons, the ground produces vegetation that by its seeds perpetuates itself according to kind, fish and birds reproduce according to kind to fill the seas and the air, animals and man do the same to fill the terrestrial realm, and from the produce of the earth animals and man receive their sustenance. Ps. 104 similarly views the creation as one functionally integrated whole.

The term "economy" is deliberately chosen in view of the kingdom metaphor. It encompasses all that pertains to the ordering, management, and function of all components within a "political" system, thus referring to more than just the "economic" aspects of such a system. The currently popular term *ecology* would not serve as well here because it would evoke a biological or more broadly organic metaphor.

76. The term used is *radah,* "rule over," as a king over a conquered region (1 Kings 4:24; Ps. 110:2; Isa. 14:2; cf. Num. 24:19; Ps. 72:8), as officials, priests, or foremen over

over the fish of the sea and over the birds of the air and over the animals, even over all the earth and over all that moves about on the earth" (1:26).[77] Like a break in syntax, this announcement effects a pause in the rhythmic progress of God's "Let there be's . . . ," jolting the reader to new attentiveness. But it is the announcement itself that dramatically illumines the event, for it is an *announcement,* not a hesitant suggestion or proposal. God announces to the members of his heavenly court, "Let us make humankind, in our image, after our likeness, and let them rule." This pause, this announcement, signalizes the creation of humankind as the most momentous of all God's creative acts. Because it entails the participation of an earthling in the office of ruling, it is fraught with great consequences for the whole divine undertaking.

And when he had created humankind in his image, he "blessed them and said to them, 'Be fruitful and increase in number; fill the earth and subdue it.[78] Rule over the fish of the sea and the birds of the air and over every living creature that moves on the ground'" (1:28). This word addressed to humankind is not a directive or commission, as long and still widely supposed. It is a benediction.[79] Having created humankind, God blessed them as he had blessed the living creatures of the air and the sea (v. 22). He blessed them in their powers of procreation and in their office of dominion over his creation; by his blessing they will fill the earth, and by his blessing they will rule over the creatures about them.

---

subordinates (1 Kings 5:16; 9:23; Jer. 5:31; Ezek. 34:4), as powerful nations over subject nations (Neh. 9:28; Ezek. 29:15), as masters over servants (Lev. 25:43, 46, 53), or as captors over captives (Lev. 26:17; Isa. 14:2).

77. Here, as elsewhere, "earth" refers not to planet Earth but to land in distinction from sky and seas. It is noteworthy that just as the creation is crowned with life, so man is viewed first of all as the steward of life in the creation—he is to rule so as to bless life.

78. The force of "subdue" *(kabaš)* has sometimes been thought to connote here an overcoming of what is hostile, rebellious, or recalcitrant. Nothing in the context suggests that man is confronted with anything of this sort in the "good" creation. Furthermore, in some contexts *kabaš* clearly does not involve overcoming, as when persons are "brought into service" in lieu of defaulted debts (cf. Neh. 5:5; Jer. 34:11, 16). A better translation here would be "and bring it into service."

79. The evidence for this comes mainly from Genesis itself and the consistent pattern of the benediction formula. First, it makes no real difference whether the introduction is "Blessed, saying . . . ," as in 1:22, or "Blessed and said . . . ," as in v. 28 (cf. Gen. 48:15 with 48:20; the usual form is "Blessed and said . . . ," as in 9:1; 24:60; 27:27; 35:9, 11). A different formula occurs if what follows is not the content of the benediction: "And he blessed him and commanded him and said to him . . ." (Gen. 28:1; cf. v. 6), "and he blessed them and he said to them, saying . . ." (Josh. 22:77). Second, the verb form of the benediction is normally imperative or jussive, as in Gen. 1:22; 9:1; 24:60; 27:27; 35:11; Num. 6:24-26 (cf. especially Jacob's recollection in Gen. 48:3-4 of the benediction received in 35:11).

According to this word, humankind as steward of the creation is the crowning component within the economy of God's visible kingdom. And according to this word, whenever humankind generates itself and whenever it advances in its understanding of the world or controls and manipulates it, God's primeval benediction is at work.[80]

This is not the place to attempt a full elucidation of what it means that humankind is created "in his image, according to his likeness." The literature is vast, and attempts at elaboration have led to various conclusions. It is at least evident, however, that this characterization is not incidental to humanity's office within the economy of the creation. Indeed, there is much to commend the reading "Let us make humankind in our image, according to our likeness, *so that* they may rule." The context suggests purpose (having created the world as his visible kingdom, God places humanity in it as the on-site representative of his rule), and the Hebrew grammar represented by "Let us . . . so that . . ." is common enough. Such a reading is further suggested by the practice of ancient Near Eastern kings to signalize their sovereignty over an acquired region by setting up in it a statue of themselves.[81] Thus the creation of humanity makes the economy of God's visible kingdom complete: it has received a creaturely representative of the Creator-King, one "formed . . . from the dust of the ground" (2:7) and yet fashioned "in the image of God."

The implications of this biblical account of God's creative acts by which "in the beginning" he founded his visible kingdom with its own integral and integrating economy are inexhaustibly rich[82] and undergird the whole biblical revelation. Here we single out only certain implications that are of special relevance for the interface between science and theology.[83] We focus first on the creaturely realm within the economy of which humanity was created to serve as God's steward, then on humankind as royal steward of the creation. For the sake of brevity, we present our observations as a series of propositions with only minimal elaboration.

80. This is not to say that God approves of all that man does with and to the creation. The greatness of man's sin is that he utilizes the growing power that is his by divine benediction to exploit and do violence to God's creatures.

81. For the grammar, see Exod. 10:2; Num. 13:2; 14:4; Judg. 16:26; 1 Sam. 17:8; 2 Sam. 18:19; etc. For the practice of Near Eastern kings, see note 44.

82. Von Rad's claim is not extravagant: "These sentences cannot be easily overinterpreted theologically!" (*Genesis*, p. 46).

83. Though theology may also be viewed as a "science," we employ the popular distinction for convenience. By "science" we mean here the empirical sciences, which differ from theology both in the objects they investigate and the methodology they employ.

The Creation as God's Kingdom

We can assert the following about the realm (God's visible kingdom) within which humankind has been created to be God's royal steward:

1. *It is real.* As that which has been called into being (which, according to alternative creation language, has been "established," "founded"), it has authentic existence—tangible, visible, audible existence. Any purely "spiritualist" metaphysics is excluded, as is any notion that the material world is merely illusory.
2. *It is intelligible.* What is *created* has been *thought;* it is the product of the mind of the Creator. And as that over which humanity has been appointed steward, it is intelligible also to the human mind. This is not to say that its intelligibility is exhaustively open to human intelligence or that a human can think the very thoughts of God concerning his creatures. But God's creatures are sufficiently knowable by the creature fashioned in his image that stewardship over them can be committed to him.[84]
3. *It is exclusively creaturely.* It is in no sense divine or semidivine, nor does it contain a divine element within it. However pervasive an exponent of God's glory it may be (cf. Ps. 19:1-4; 29:9; 104:1), even the powers at work in it, whether atomic, gravitational, biological, or intellectual, are wholly other than divine.[85]
4. *It exhaustively encompasses all that belongs to the visible world.* The small and great, from the least of the seeds (and pollen) of the plants and the least of the forms of life in the seas and on land to the firmament, the distant stars, and the oceans above and below—all things in the physical universe belong equally to the realm that is governed by the creation decrees of God and the God of the creation decrees. The specifically named entities in Genesis

84. For these two implications see also Daniel O'Connor and Francis Oakley's introduction to *Creation: The Impact of an Idea* (New York: Charles Scribner's Sons, 1969), pp. 7-9.

85. This world is thus radically different from that of the ancient Near Eastern myths in that it does not share a primordial substrate with the gods. It does not even contain within it any inherently sacred places (such as mountains or "bethels," sites frequented by the gods [cf. Gen. 28:16-19]), persons (such as kings, priests, or shamans), or forces (such as sexual generation) that participate in the quality of holiness. Even such linkage of the created with the divine as is posited in various forms of the Aristotelian chain of being is excluded. Though all creation is revelation of God and sacred to him as "the works of his hands," all that is created is not-god, is equally not-god. See further Hans Jonas, "Jewish and Christian Elements in the Western Philosophical Tradition," in *Creation: The Impact of an Idea,* pp. 245-46, for a list of eight characteristic aspects of Greek philosophical thought about the world that differ from the biblical.

1 reflect the observational phenomenology of the ancient world, but its all-inclusive scope leaves no doubt that however vast the cosmic dimensions and however minute the smallest particles of matter, all are declared to be under the rule of the Creator-King.

5. *It possesses its own integral and integrating economy.* Each of its components has its own internal economy (e.g., the biological economy of plants and animals), and all its components were created to fill out and integrate the economy of the whole. It is not a mere aggregate of disparate kinds and orders of being that humanity (as steward) has to subdue into harmony. Nor is it a mix of forces and functions that exist independently and unrelatedly.

Furthermore, the internal economy of the created realm is neither incomplete nor defective.[86] That is to say, it contains no gaps that have to be filled by continuous or sporadic *immediate* operations of divine power;[87] God is not himself a component within the internal economy of his creaturely realm.[88] Nor is the evil that is in the world the result of defect in the economy of the creation. All that is rightly judged to be evil in the world—all that is evil relative to God's will for and his purposes embodied in the creation—is introduced with the fall. But even that evil is limited to such as is consonant with the economy of the creation. Sin itself, as moral evil, is conditioned by the creation order. Neither the sin that entered the world with the fall nor the divine curse and judgments that followed have fundamentally altered the creation or its economy.[89] The divine curse, just as the divine blessing, is

86. This is underscored in the creation account by the sevenfold divine assessment that the creation is "good" and by the divine "rest" (ceasing from creation because the work was finished).

87. The one notable exception often posited by Christian theologians is the creation *ex nihilo* of individual human "souls." The large issues involved cannot all be explored here, but we do note the following: (a) Christian speculation concerning the origin of individual "souls" has led to no consensus; (b) the biblical view that humanity reproduces itself by divine blessing of its reproductive powers and the biblical statement that Adam "begot a son in his own likeness, after his own image" (Gen. 5:3) hardly suggest a dual source of the human individual, one "natural" and the other "miraculous"; and (c) even if the latter is the case, it has no bearing on those aspects of the creation that are investigated by the natural sciences.

88. This is not to say that the economy of the created realm is self-sustaining, that it operates independently of the creator's sustaining and governing sovereignty.

89. The age-old view that death entered the biological realm only with the postfall curse is neither affirmed or implied in Genesis. At best it can be said of it that it is speculative extrapolation from figurative biblical depictions of eschatological conditions. Even man, according to Gen. 3, would have "lived forever" only if he had retained access to "the tree of life."

relative to the economy of the creation. The claim that the second law of thermodynamics is a result of God's primeval curse having effected a fundamental alteration in the very nature of physical reality is groundless speculation void of any biblical warrant.

6. *It is pervasively contingent.* Neither the specificity of its components nor the economy that integrates them is generated by *necessity;* rather, they are products of the free will of the Creator. Moreover, the outworking of the economy of the creation, the actual course of creation's history, is not controlled by some locked-in inevitability. It too is contingent, subject to the overarching and all-pervasive rule of God. The biblical metaphor for the creation is that of kingdom, not machine, and as such it is always and everywhere open to the directing rule of its Creator-King. Moreover, God's creation decrees, his "Let there be's . . ." are accompanied by divine benedictions that are effectual not automatically but only as God wills. In addition, the economy of the created realm includes the stewardship of a creature created "in the image of God" who possesses freedom and bears moral responsibility for the execution of his ruling office. A realm containing within its economy such a component is open to the effects of free choices on the part of the steward as well as the free decisions on the part of the Creator-King in his dealings with his creation.
7. *It is vulnerable to human misconstrual and exploitation.* As a reality that has been given an existence distinctly separate from (though not independent of) God, a realm that is intelligible and that possesses its own internal integrated economy, the creation is open to serious misconstrual. Eyes that can see it, ears that can hear it, noses that can smell it, hands that can feel and handle it, minds that can understand it, can conceive of it as self-contained (either as a plethora of gods or as Nature writ large—an organism, a machine, a self-directed or chance-guided complex of forces). Its dimensions, substances, qualities, mechanics, forces, functions, activities, regularities, and inner dynamic relationships can be investigated and discovered without knowledge of the Creator. And to the degree that *these aspects* of the creation are understood, it can be acted upon, reshaped within its given qualities and economy, and exploited by the creature who has been given power to "rule over" it. The steward within the economy of the visible kingdom of God can exercise control over the creation even in alienation and rebellion.

Humankind as God's Royal Steward

We can assert the following about humankind as created "in the image of God" so that it may exercise a steward's dominion over "the works of God's hands":

1. *It is a component within the economy of the created realm as God's royal steward.* Though fashioned in God's image, humans are "of the dust of the ground." Our function is unique, we are gifted with unique powers consonant with that function, and we stand in a unique relationship with the world about us, but we are not alien beings in the realm in which we are stewards. Nor is stewardship of the creation incidental to our "essential" identity; ours is not a "personhood" to which stewardship has been given as a temporary or probationary assignment. Stewardship belongs to our being—as governing to a king and illumination to a lamp.
2. *It is a steward whose dominion is limited to the visible created realm.* Humankind has no power to rule over the Creator or over the creatures that serve in God's heavenly court. Though we may appeal to God, we cannot manipulate him; though we are ministered to by the heavenly host, we cannot command them.[90]
3. *It is a steward whose dominion is limited to "ruling over" the creatures.* Unlike God, we cannot create fundamentally new components within the created world. Unlike God, we cannot maintain the existence of the created world. Unlike God, we cannot alter the economy of the created world. We can only, by working within the economy of the created world, bring the creatures into service: we can only "cultivate and take care of the garden" (2:15). This includes, of course, the ability to rearrange existing components into new materials and structures, such as high-temperature superconductors and genetically altered species.
4. *It is a steward whose dominion over "the works of God's hands" presupposes a capacity to understand the creation.* Ours is a kingly office, and to rule requires authentic insight into the realm to be governed.[91]

90. The author of Hebrews sums up the role of the heavenly host as follows: "Are not all angels ministering spirits sent to serve those who will inherit salvation?" (1:14). They are "sent" by God and they "serve" mankind only as they minister on God's behalf to the saints.

91. Accordingly Solomon asked for and received wisdom to govern Israel (1 Kings

5. *It is a steward whose insights into the creation are relative to its place and office within the economy of the visible creation.* Such limitation is entailed in our creatureliness, but also in the office for which we were created. As the biblical authors frequently remind us, God's wisdom transcends that of any human being (Job 28; 38:37; Ps. 147:5; Prov. 8:22-31; 25:2; Isa. 40:13-14, 28; Rom. 11:33-36; 1 Cor. 13:9, 12).
6. *It is a steward whose insights into the created realm are gained through experience of the creation.* Ours is not an innate knowledge. God *brought* the animals to Adam to see what he would name them and to see if he would recognize among them a "helper fit for him" (2:19), and he brought the woman to Adam for the same purpose (2:22).
7. *It is a steward whose insight into the created realm is progressive as its experience of the creation is extended and expanded,* also through the generations. Humanity's dominion over creation and efforts to bring it into service are not complete in the beginning but progress through the continuing effectiveness of God's primeval benediction—as God chooses to effect it.
8. *It is a steward that both maintains itself and serves its Creator-King by the exercise of its dominion over the creation.* We are dependent on and assigned sustenance from the very realm over which we have been appointed stewards. (This inseparable relationship between our office and our sustenance opens to us the possibility—which in history we have actualized—of dreaming of independence from our Creator and consequently engaging in massive exploitation of the creation.)
9. *It is a steward whose insights into and rule over the created realm can*

---

3); Messiah would be gifted with wisdom to rule justly (Isa. 11:2); and Lady Wisdom is heard to say:

> By me kings reign
> and rulers make laws that are just;
> by me princes govern,
> and all nobles who rule on earth. (Prov. 8:15-16)

The first "Solomonic" proverb recorded by Hezekiah's schoolmen puts it succinctly: "It is the glory of God to conceal a matter"—his wisdom transcends man's; "to search out a matter is the glory of kings"—governing requires insight (Prov. 25:2). In fact, the central theme of Proverbs is that humans can do nothing effectively without wisdom (*ḥokmah,* which includes knowledge, insight, and understanding). Even expertise in technology is afforded by "wisdom" (Exod. 31:3-6), as is competence in agriculture (Isa. 28:23-29) and the ability to speak encyclopedically of the array of the creaturely realm (1 Kings 4:29-34; cf. Prov. 30:15-31).

be effective and progressive even when it is alienated from the Creator. This, too, is entailed in our being created "in the image of God" and in our place within the economy of the creation realm. It is for this reason that after Adam was expelled from the Garden he could yet cultivate the barren earth and wrestle from it the nourishment needed to sustain him (3:23). Accordingly, Cain developed agriculture as effectively as Abel developed the pastoral arts (4:2); he developed the crafts of building (4:17), and his descendants developed nomadic pastoralism as well as elementary musicology and metallurgy (4:20-22). Moses received his basic education about the world from the Egyptians (Acts 7:22). The Israelites learned agriculture and horticulture from the Canaanites. Solomon imported craftsmen from Phoenicia for his great building projects (1 Kings 5:6; 7:13-14) and drew on that same people to sail and navigate his ships (1 Kings 9:27). Whatever the misconstruals and distortions resulting from humanity's alienation from God, the fall did not diminish human noetic powers[92] or cut humanity off from effective experience of the world. Human understanding of the creation has continued to grow apace, however much people have distorted their perceptions of the Creator—whether in the ancient manner of viewing the powers within the creation as themselves divine or in the modern way of conceiving of the creation as self-creating Nature. Scientists may at times be mischievous—as even theologians may be mischievous—but humanity's growth in understanding of the creation and its ability to exercise dominion over the creation is undeniable and, according to Genesis 1, is to be credited to the continuing effectuation of God's benediction "in the beginning."

Implications for Science and Theology

These observations concerning the created realm as God's visible kingdom and mankind's role within its economy as God's royal steward

92. Nowhere does Scripture speak of humanity's *diminished* intellectual capacities or bemoan its loss of mental power. That there are limits to humankind's knowledge is often acknowledged, but these are due to humanity's finiteness, not to its fallenness. However much the blinded fools humans may become, suppressing knowledge of God (Rom. 1:18-22) and failing to discern the right way or moral truth, the cause is not a disabling imbecility but a corrupted will, a culpable alienation from God and concomitant moral disorientation.

entail much for both science and theology and for their relationship. We cannot unwrap all their implications here. For our purposes we have singled out those that appear to us most pertinent.

Fundamental to all else is the fact that human knowledge of God comes *with* knowledge of the world and not apart from it.[93] As creatures in the world, ourselves a component (as God's stewards) within the economy of the creation realm, we know God in his role as Creator of the world. Our knowledge of God is not innate, nor is God's revelation of himself to us immediate—infinite Mind in immediate communion with finite mind (as certain forms of idealism, mysticism, and personalism would have it). As Creator, God discloses himself to his creature in the context and through the display of his glory in the creation. All God's dealing with humankind, including his revelation of himself, are with the creature "in his image" who has been created an integral part of and a "dominion"-wielder in the creation. A servant knows his lord only within the frame of reference of the lord's estate and his own place within the economy of that estate. Paul agrees with this understanding of the character of the human knowledge of God: "What may be known about God is plain to them, because God has made it plain to them. For since the creation of the world God's invisible qualities—his eternal power and divine nature—have been clearly seen, being understood from what has been made, so that men are without excuse" (Rom. 1:19-20). His word to the Athenians is to the same effect (Acts 17:24-28). In the language of theologians, special revelation presupposes general revelation.[94]

It is equally true, of course, that we do not truly know ourselves or the world apart from our knowledge of the Creator. To know the world truly is to know it as creation, as the visible kingdom of the

93. Compare von Rad's assertion relative to the Old Testament view of "wisdom": "The experiences of the world were for her [Israel] always divine experiences as well, and the experiences of God were for her experiences of the world" (*Wisdom in Israel,* trans. James D. Martin [Nashville: Abingdon, 1972], p. 62).

94. On this Calvin is emphatic:

> God—by other means invisible— . . . clothes himself, so to speak, with the image of the world, in which he would present himself to our contemplation. They who will not deign to behold him there magnificently arrayed in the incomparable vesture of the heavens and the earth, afterwards suffer the just punishment of their proud contempt in their ravings. Therefore, as soon as the name of God sounds in our ears, or the thought of him occurs to our minds, let us clothe him with this most beautiful ornament; finally, let the world become our school if we desire rightly to know God. (Prefatory argument to *Commentaries on the Book of Genesis,* trans. John King [1947; reprint ed., Grand Rapids: Eerdmans, 1948], p. 60.)

Creator; to know ourselves truly is to acknowledge that we are the Creator-King's stewards in the creation. "The fear of the LORD is the beginning of wisdom" (Ps. 111:10; Prov. 9:10). True knowledge involves knowledge of God, humanity, and world in their interrelationships. Accordingly, every claim to knowledge implicitly involves a whole metaphysical system.

God's royal steward in the creation faces both God and the world. We face God as those who have been given the power and responsibility of dominion in the world; we face the world as those who are responsible for representing God's righteous and benevolent rule in the world. We thus serve as mediators between God and world. And just as prophets, priests, and kings were chosen from among the Israelites to mediate God's rule over his people (Deut. 17:15; 18:18; Heb. 5:1), so we as mediators of God's rule in the creation are ourselves components within the economy of the creation. As those formed "from the dust of the ground" and dependent on the fruit of the ground, we are made "in the image of God," and by virtue of this twofold reference we are God's stewards in the visible creation.

This mediatorial stewardship defines humanity's vocation. As we face the world, we must do so as those who know the Creator-King; as we face God, we must do so as those who know the creation. We can fulfill this vocation, fulfill the very purpose of our being, only as we rightly know both God and the creation. (To know rightly does not mean, of course, to know exhaustively or to know as God knows. It is to know God, self, and world so as to be faithful and effective stewards.) Thus both theology and science belong to humankind's vocation (though, to be sure, theology is only a secondary and limited aid to human knowledge of God, and science is only a secondary and limited aid to human knowledge of the world: theology assumes an antecedent knowledge of God, and science assumes an antecedent knowledge of the world).

And in view of the relationship between God, humanity, and world (Creator, steward, and visible kingdom), of the nature of divine revelation (special revelation presupposing general revelation), and of the role of humankind as stewards of the creation, these two aspects of humanity's vocation—theology and science (a deliberate and disciplined pursuit of knowledge of God and world)—cannot be pursued in isolation from each other. Theology must take account of all that humanity comes to know about the world, and science must equally take account of all that we come to know about God. In fact, we cannot, without denying our being and vocation as stewards, pursue theology without bringing to that study all that we know about the world, nor can we, without denying our being and vocation as stewards, pursue science without bringing to that study all that we know about God. Thus the intercourse between theology and science must be continuous. It must also retain the character of an authentic dialogue: neither one may assume an attitude of superiority to the other.

Regarding human knowledge of the creation, we further observe the following:

1. In view of the pervasive contingency of the creation, we cannot know it by pure speculation about the world. Nor can we know it by rationalistic elaboration of first principles, after the manner of ancient and early modern speculative philosophers. *For that reason, empirical science as a disciplined method for directing and testing our experience of the world is a more appropriate means for the fulfillment of our vocation as royal stewards of the creation.*
2. Since the created realm is replete with its own economy that is neither incomplete (God is not a component within it) nor defective, *in our understanding of the economy of that realm so as to exercise our stewardship over it*—understanding based on both practical experience and scientific endeavors—*we must methodologically exclude all notions of immediate divine causality.*[95] As stewards of the creation we must methodologically honor the principle that creation interprets creation; indeed, we must honor that principle as "religiously" as the theologian must honor the principle that "Scripture interprets Scripture"—or, since Scripture presupposes general revelation, that revelation interprets revelation. In pursuit of a stewardly understanding of the creation, we may not introduce a "God of the gaps," not even in the as-yet mysterious realm of subatomic particles. We may not do so (1) because God is not an internal component within the economy of the created realm, and (2) because to do so would be to presume to exercise power over God—the presumptuous folly of those in many cultures who have claimed to be specialists in the manipulation of divine powers (e.g., shamans in Russian folk religion and medicine men in primitive cultures).
3. As an intelligible realm that displays the glory of the Creator and

95. This is not to deny the reality of biblically attested miracles but rather to assert (a) that the miraculous does not have a *normal* role in the economy of the creation, (b) that acceptance of biblical witness to miracles does not contribute to knowledge of the *economy* of the creation, and (c) that miracles do not play any part in our exercise of day-by-day stewardship in the creation.

as a realm over which humanity has been appointed God's royal steward, the creation contains no inner deceptions. Humans can misread the phenomena, misinterpret their experience of the world, introduce distortions in their gathering of data, pursue misguided research, and employ wrongheaded principles of explanation, but the creation itself does not mislead. *As God's appointed stewards over God's creation, we can trust the integrity of compelling evidence to lead us into a valid understanding of the creation.* The Creator is not a deceiver—how then could the creation display his glory? And his creation does not mislead—how then could God's royal steward fulfill his vocation?

CONTEMPORARY SCIENTIFIC COSMOGONY AND THE BIBLICAL DOCTRINE OF CREATION

We have undertaken this discussion of the biblical doctrine of creation to assist the Christian community in assessing the current scientific reconstructions of cosmic history.

On the face of it, the biblical and scientific accounts of cosmic origins appear to stand in radical opposition. As a consequence many have felt compelled to choose between them. Some, convinced by the probative force of the mass of accumulated data, have dismissed the biblical word as the misguided religious imaginations of ancient Israelites. Others, out of a simple faith that Genesis 1:1–2:3 is God's straightforward record of the history of the world's formation, have passionately rejected the cosmogonic reconstructions of the scientists as the false speculations of atheistic minds. But as we have seen, there have also been those who with great wrestling of soul have participated in the scientific study of the creation while holding firmly to their faith that the Bible (including Gen. 1:1–2:3) addresses them as the Word of God. Though often viewed by those to their right as faithless compromisers and by those to their left as self-deceived fools, they have held staunchly to the conviction that God has not deceived them either in his creation or in his Word and that the study of both must go hand in hand. They have learned—and it has humbled them—that human understanding of the Bible is as subject to fault as human understanding of the creation.

It is our conviction that these last have chosen the better way, that their way is most faithful to Genesis 1:1–2:3 itself. That word was not intended as a chronicled account of the origins of the cosmos. It speaks of beginnings, but in a manner quite different from that of the historian or scientist. It is a theological word written in the face of ancient Near Eastern myths with their theogonies and comogonies. It offers a radically new view of God, humanity, and world that funds the whole biblical witness to God's pursuit of his purposes in history.

From this word concerning God, humanity, and world comes the charter for our pursuit of knowledge about the world, including the development of science. Science belongs to humanity's vocation as God's royal steward of the creation. That this pursuit of knowledge has not been fruitless is grounded in our human capacities as God's image-bearers and stewards, the intelligibility and integrity of the creation, and the continuing effectuation of God's benediction upon his steward.

To be faithful to this word, Christians must take science seriously as belonging to their stewardship and must take seriously what science discovers about the world. As the great Christian theologian Calvin once wrote,

> if the Lord has willed that we be helped in physics, dialectic, mathematics, and other like disciplines, by the work and ministry of the ungodly, let us use this assistance. For if we neglect God's gift freely offered in these arts, we ought to suffer just punishment for our sloth.[96]

Accordingly, in his commentary on Genesis 1:16, after observing that Moses called the Sun and Moon "the two great lights" but that astronomers had proved Saturn to be larger than the Moon, he adds this pastoral word:

> Nevertheless, this study is not to be repudiated, nor this science to be condemned, because some frantic persons are wont boldly to reject whatever is unknown to them. For astronomy is not only pleasant, but also very useful to be known: it cannot be denied that this art unfolds the admirable wisdom of God. Wherefore, as ingenious men are to be honoured who have expended useful labour on this subject, so they who have leisure and capacity ought not to neglect this kind of exercise.[97]

96. Calvin was deeply convinced that the Lord did so will, insisting that truth about the world, even if brought to light by the ungodly, is a "benefit of the divine Spirit," *Institutes*, 2.11.16.

97. Calvin, *Commentaries on the Book of Genesis*, vol. 1, pp. 86-87. Elsewhere he comments, "He who would learn astronomy, and other recondite [abstruse] arts, let him go elsewhere [than to Gen. 1]." (p. 79). See also Calvin's assessment of the sciences in "A Warning against Judiciary Astrology and Other Prevalent Curiosities," trans. Mary Potter, *Calvin Theological Journal* 18 (1983): 156-89.

Science is, of course, a fallible venture, and it has tended in recent years toward a greater modesty in its claims. Still, its proven successes relative to "inferior things" (the term is Calvin's) ought to constrain the Christian not to dismiss out of hand what natural science has brought to light or those interpretations of the world that science offers on the basis of many interlocking bodies of empirical data.

Specifically, since the Bible provides no *detailed record* of the genesis of the physical creation but does provide a charter for humanity's scientific endeavors, Christians ought to accept, with the tentativeness due to all scientific claims, the current scientific reconstructions of the progressive cosmic and planetary development. For them the process by which the present world has come to be—a process of unimaginable dimensions and intricacy, of which the traces meet our eyes at every turn—ought to evoke a sense of overwhelming wonder at "the admirable wisdom of God." If this be the visible kingdom of God and this its marvelously intricate and integrating economy, then, standing in this temple of the Creator-King, we cannot but join all creatures and cry "Glory!" (Ps. 29:9). With eyes that have been enabled to see in the creation more than the ancients saw, we can take up their song fortissimo:

> O LORD my God, you are magnificent;
> you are clothed with spendor majestic. (Ps. 104:1)

But even so, we do not and cannot yet sing as those who chorus the praise of the Creator on the other side of the eschatological disclosure of God's glory. We see as we can see and we praise as we can praise in the midst of history, where knowledge remains "in part" (1 Cor. 13:9). We do not yet understand the creation as we shall understand it; we do not yet understand the Scriptures as we shall understand them. We remain stewards of the creation and stewards of "the . . . words of God" (Rom. 3:2). Theology and science remain for us unfinished tasks. Not that faith and service await the perfection of knowledge. For that reason, what Paul wrote to the Philippians in a somewhat different context is apropos here: "Only let us live up to what we have already attained" (3:16).

As the history of Christian thought has abundantly shown, the struggle to understand the Scriptures in the light of new knowledge of the world and the attempt to understand the world in the light of new knowledge of the Scriptures is sometimes an agonizing business. To have to reassess views that have long been held as firm (because unquestioned) can be unsettling. From church history comes the classic example of the painful reorientation in both biblical interpretation and dogmatic constructions required by the Copernican-Galilean revolution in our understanding of the solar system. Now that that struggle is over and memory of its pain has faded, we can no longer appreciate its intensity or why it should have been so passionate and so prolonged. But we would do well to recall such episodes in the history of Christian thought in order to face with patience and greater equanimity the issues confronting us today.

8. EPILOGUE

Where Do We Go from Here?

HOWARD J. VAN TILL

IN THE PRECEDING CHAPTERS we have dealt at length with several questions concerning the character of both biblical and scientific portraits of the Creator's handiwork. Perhaps in some ideal world well-informed Christians would have little difficulty accepting each of these as a valid portrait and would easily recognize the need for honoring the ways in which these portraits differ from one another as a consequence of different choices of standpoint.

The situation might be comparable in part to one encountered in photography. If, for example, we wished to photograph a house, we might choose to do so from several different standpoints, obtaining a front view as seen from the street, a side view as seen from a neighboring yard, an aerial view as seen from a hot-air balloon passing overhead. The several photographs taken would differ substantially from one another, but we would certainly not insist that they be declared inconsistent or contradictory. Neither would we say that only one of them could be true and that the others must be rejected as false. Rather, recognizing how the choice of standpoint imposes limits on what any one photograph is able to portray, we would, in fact, find it helpful to have several, thereby providing us with a more comprehensive perspective on the appearance of the house. And we might even supplement the photographs with architectural drawings or construction blueprints to supply information not visible in a photograph.

But the actual project of comparing scientific and biblical portraits of the universe is not so simple a matter, and the task of incorporating both to achieve a more complete understanding of God's handiwork seems at present to be proceeding very awkwardly. Unhappily, the warfare metaphor too frequently provides the colorful and militant vocabulary with which popular discussions concerning the relationship of natural science to Christian belief are conducted. In place of rational and amicable dialogue directed toward the goal of mutual understanding and communal wrestling with complex issues, we too often find heated and antagonistic debate dedicated to defeating opponents and developing ideological factions within the Christian community.

Having experienced both the joy of Christian encouragement and the sting of hostile reaction, we reflect in this brief epilogue on some of the roots of today's religiously motivated resentment of science and offer a few modest suggestions concerning what Christian leaders might do to improve the relationship between scientists and nonscientists within the Christian community.

WHY DOES THE WARFARE METAPHOR PERSIST?

Why is it that "science" is so often perceived by Christians today as if it were a hostile, faith-threatening agent to be resisted or rejected? While a substantial part of the blame for this state of affairs must be borne by outspoken proponents of naturalism (we'll return to this aspect shortly), some of it falls on the shoulders of the Christian community itself. The character of our Christian training (in some cases indoctrination) may inadvertently (in some cases intentionally) encourage many within the Christian fold to develop a deep suspicion of and resentment toward the scientific enterprise.

In the normal course of our education we are all presented with "packets" of several concepts that appear to be bundled together because they are given to us by the same kind of people. Especially in our formative years we receive from authority figures—parents, teachers, preachers—instruction in a diversity of areas: practical skills, knowledge in the academic disciplines, a world picture, biblical knowledge, theological concepts, and so on. But these highly varied instructions may not be very well differentiated in terms of their relative importance within a Christian worldview or concerning the degree of certainty with which specific concepts should be held. Then, especially if our vocations or careers do not involve us directly in assessing the individual components of our "perspective packet," we may easily be led to think that the entire packet stands or falls *as a complete unit*.

When that is the case, a challenge to any one component in the packet is perceived as being a threat to the entire system. We have all heard the rhetoric: "If God did not create the world in six literal days as he says he did in Genesis 1, then the whole Bible must be false and God must be a liar"; and, "If there was no literal worldwide flood just a few thousand years ago and within human history, then we can no longer have confidence in any biblical reference to historical events"; and, "If we do not see God's work of creation in terms of instantaneous miracle, then we must reject all of the miracles reported in the Bible"; and, "If the physical death of all creatures is not the consequence of the fall in the Garden of Eden, then there is no salvation in Christ's victory over death in his resurrection."

In the context of such an "all-or-nothing" approach, a call for change will likely be perceived not merely as intellectually incorrect on some specific matter but as profoundly unfaithful to the entire received tradition. The response to such an alarming state of affairs is typically bold and firm: protect the received tradition packet *at all costs;* repudiate all innovation by labeling it "heresy," "apostasy," or "blasphemy."

One significant component in our received tradition packet is a world picture. (Recall our introduction of the worldview / world picture distinction in chap. 2, note 3.) But in the early and medieval Christian tradition the prevailing static world picture (drawn largely from early Greek thought) was generally translated into a "creation picture" in which God not only called the whole creation into existence at the beginning but also immediately—both in the sense of without delay and without the employment of secondary causes—gave final and essentially fixed form to each basic structure and each "kind" of creature. Within a week of the beginning, about six thousand years ago in this creation picture, there were spiral galaxies, our solar system, planet Earth, butterflies, dinosaurs, and human beings.

However, during the past couple of centuries, scientists have undertaken a careful *empirical* study of the world's formative history and have concluded that it is marked by more or less continuous processes of formation—all consistent with certain fundamental "natural" patterns of material behavior—leading historically to a diversification of both physical structures and life-forms over a fifteen-billion-year period. The present life-forms, it is believed, are genealogically related to past forms but also display an interesting degree of novelty relative to their ancestors.

The contemporary state of affairs practically ensures a conflict between religiously held tradition and the current conclusions of natural science. The traditional world picture and its attendant "creation picture" are tightly linked with the Christian worldview and the "faith of the fathers." In some cases (e.g., North American fundamentalism) the authenticity of the entire belief system is thought to be wholly dependent upon the factual accuracy of the received creation picture born from the marriage of the static world picture of Greek philosophical tradition and a literalistic interpretation of the early chapters of Genesis.

But the modern, scientifically informed world picture is substantially different from the one embedded in the received "faith packet." Hence, modern science is viewed by many as the faith-destroying enemy of Christianity. More specifically, the words *creation* and *evolution* have come to be used as labels for irreconcilably different answers to the question of how we got here. The vigorous perpetuation of the warfare metaphor is practically assured.

The problem is further exacerbated in the United States by the characteristic features of the American educational system. The public schools are not allowed to promote a particular religion. In effect, this prohibition is often translated into the practice of excluding any reference whatsoever to theistic religion or to the important role that it has played in American culture.[1] Private Christian schools, on the other hand, tend, as extensions of parental training in Christian faith and life, to perpetuate the received tradition with minimal critical examination. When there appears to be a conflict between the "faith packet" and the conclusions of allegedly "secular" science, the temptation is to dismiss contemporary scientific scholarship as fundamentally misguided (guided, that is, by unacceptable principles and presuppositions) and consequently not worthy of a sympathetic hearing or of open evaluation. Christian educators often feel caught on the horns of a dilemma in which either professional standards or parental approval may have to be sacrificed.

For a large portion of the Christian community the net result is a loss of valuable interaction between the activity of scientific scholarship and the articulation of Christian belief. For many this loss of interaction leads to a diminishing ability to incorporate the results of natural science into a contemporary expression of the historic Christian faith. Others experience a declining appreciation of the relevance of Christian faith to the scientific enterprise or a reduced awareness of the territorial limits of the scientific world picture. And some have recounted to us their soul-wrenching experience of feeling compelled to make an either-or choice between Christian faith and intellectual honesty.

1. For a discussion of the way in which this has affected public school textbook material, see Paul C. Vitz, *Censorship: Evidence of Bias in Our Children's Textbooks* (Ann Arbor: Servant Books, 1986).

But the persistence of the warfare metaphor is not the fault of Christians alone. Numerous advocates of the worldview of naturalism must share the blame. With singularly undeserved success they have indentured the results of the scientific enterprise in the service of providing the appearance of warrant for their atheistic worldview. The literature of popularized science provides many examples of this familiar rhetoric.

Isaac Asimov, the most prolific of popular science writers, asserts, for example, that "In short, the scientific view sees the Universe as following its own rules blindly, without either interference or direction."[2] In essence, Asimov is here making the standard naturalistic leap of illogic by presuming that if natural science (which, as we pointed out in chap. 5, is *methodologically* blind to the reality of divine governance) does not directly perceive God to be governing and directing his Creation, then the universe must be autonomous and undirected.[3] That's equivalent to the case of a person listening to a symphony on the radio asserting that, since she cannot personally see the conductor, the orchestra must be playing without direction.

In a short book inappropriately entitled *The Creation,* British chemist P. W. Atkins boldly announces, "My aim is to argue that the universe came into existence without intervention, and that there is no *need* to invoke the idea of a Supreme Being."[4] Later, following a maze of speculative rhetoric, he claims that "we have been back to the time before time, and have tracked the infinitely lazy creator to his lair (he is, of course, not there)."[5] And summarizing his argument, presented as if it were a product of *scientific* theorizing, he says, "In a word, the central speculation is that space-time generates its own dust in the process of self-assembly. The universe can emerge out of nothing without intervention. By chance."[6]

Similar pronouncements of naturalism can be found in the writings of Oxford zoologist Richard Dawkins. In *The Blind Watchmaker,* for example, he caricatures the Christian concept of creation in the architec-

2. Asimov, *In The Beginning* (New York: Crown, 1981), p. 11.

3. Recall, for example, our emphasis that although natural science has the competence to study the physical behavior of matter, it does not (by its choice of instrumentation and methodology) have the ability to determine whether that behavior is self-governed or God-governed. The Christian scientist *as a whole person* may view all physical phenomena as subject to divine governance, but that is a belief that goes beyond the limited domain of scientific inquiry.

4. Atkins, *The Creation* (San Francisco: W. H. Freeman, 1981), p. vii.

5. Atkins, *The Creation,* p. 115.

6. Atkins, *The Creation,* p. 113.

tural and mechanical imagery of the watchmaker analogy made famous by the eighteenth-century theologian William Paley—an analogy rightly discarded by most contemporary Christian thinkers. Dawkins then proceeds with a flourish of triumphalist rhetoric to declare the victory of Darwinism over this straw defender of an inadequate creation picture. In a direct assault on the concept of purposefully directed divine governance, Dawkins asserts that "Natural selection, the blind, unconscious, automatic process which Darwin discovered, and which we now know is the explanation for the existence and apparently purposeful form of all life, has no purpose in mind."[7] Darwin's concept of natural selection, presented in the metaphor of the *blind* watchmaker, is persistently exploited by Dawkins to play a key role in the mythology of his evolutionary naturalism. And with a *non sequitur* calculated to inspire hostile reaction from Christians—especially from those who do not recognize its logical deficiencies—Dawkins would have us believe that "although atheism might have been *logically* tenable before Darwin, Darwin made it possible to be an intellectually fulfilled atheist."[8]

A decade ago Carl Sagan, in his public television series and book *Cosmos,* also presented science as if it provided the exclusive pathway to knowledge and as if it had the competence to expose all theistic religious beliefs as little more than brightly robed statements of ignorance. Although he is able to admit that science is an imperfect and often misused tool, he is also quick to offer it a litany of high praise: "But it's the best tool we have, self-correcting, ever-changing, *applicable to everything.* With this tool we vanquish the impossible."[9] And what are the implications of modern scientific cosmology for the theological concept of creation? According to Sagan, "If the general picture of an expanding universe and a Big Bang is correct, we must then confront still more difficult questions. What were conditions like at the time of the Big Bang? Was there a tiny universe, devoid of all matter and then the matter suddenly created from nothing? How does *that* happen? In many cultures it is customary to answer that God created the universe out of nothing. But this is mere temporizing."[10] Sagan apparently confuses the theological explanation of *origin* with the scientific description of *formation,* a common failure in contemporary popular science writing.

In a recent and remarkably popular book, *A Brief History of Time,* theoretical physicist Stephen Hawking not only provides a readable

7. Dawkins, *The Blind Watchmaker* (New York: W. W. Norton, 1986), p. 5.
8. Dawkins, *The Blind Watchmaker,* p. 6.
9. Sagan, "Cosmos," episode 12; italics added.
10. Sagan, *Cosmos* (New York: Random House, 1980), p. 257.

overview of contemporary cosmology but also presents some of his own speculations and what he perceives to be their theological implications. Specifically, Hawking speaks of "the possibility that space-time was finite but had no boundary, which means that it had no beginning, no moment of Creation."[11] Thus, when Hawking makes occasional reference to God or to a creator, those references are highly provisional. In fact, if his proposal is correct, then the need for any reference to a Creator God will have been, by his measure, eliminated: "The idea that space and time may form a closed surface without boundary also has profound implications for the role of God in the affairs of the universe. . . . So long as the universe had a beginning, we could suppose it had a creator. But if the universe is really completely self-contained, having no boundary or edge, it would have neither beginning nor end: it would simply be. What place, then, for a creator?"[12]

As Christians we have every right to be irritated, even offended, by some of the rhetoric of writers like Asimov, Atkins, Dawkins, Sagan, and Hawking.[13] Their highly provocative religious statements, presented in the cloak of natural science, are bound to push many Christians into repudiating the scientific enterprise. While insightful readers will recognize in such literature the numerous logical fallacies in argumentation and the mischievous disregard for the limited competence of natural science, many others will be misled to believe that a major goal of the scientific enterprise is to discredit the Christian faith. It would appear, in fact, that many Christians have uncritically accepted the idea, implicit in the material cited above, that if the scientific *world picture* is correct, then the naturalistic *worldview* logically follows. Nothing could be further from the truth; but once that fallacious claim is granted, the only remaining apologetic strategy for Christians would seem to be a vigorous attack on the credibility of some major element in the contemporary scientific world picture.

An extreme example of this strategy is represented by the fundamentalist "creation science" movement. Since this is discussed at length in chapter 6, we will say no more about it here.

A more moderate version, considered by several Christian scholars today to be worthy of critical examination, might be called

11. Hawking, *A Brief History of Time* (New York: Bantam Books, 1988), p. 116.

12. Hawking, *A Brief History of Time*, pp. 140-41.

13. For more extended critiques of the rhetoric of naturalism in popular scientific literature, see Howard J. Van Till, Davis A. Young, and Clarence Menninga, *Science Held Hostage: What's Wrong with Creation-Science AND Evolutionism* (Downers Grove, IL: Inter-Varsity Press, 1988), chaps. 7-9.

"old-earth special creationism."[14] For its proponents, this middle position (somewhere between "creation science" and evolutionary creationism) has several attractive features:

1. It incorporates the generally accepted scientific conclusions concerning the multibillion-year formative history of inanimate structures such as galaxies, stars, and planets;
2. it holds an important place for the concept of special creation, thought by many Christians to be required by the Bible;
3. it summarily rejects the concept of unbroken genealogical continuity, a prominent element in the contemporary scientific world picture that has been incorporated into the mythology of the naturalistic worldview; and, some would add,
4. it is consistent with the concept of an ontological hierarchy of "kinds," considered by some Christian philosophers to be biblically required.[15]

Critics of this approach would be quick to suggest the following, however:

1. It appears to be methodologically inconsistent by allowing for continuity in the formative history of inanimate structures on the basis of scientific findings in spite of Genesis 1 while insisting on discontinuity in the genealogy of life-forms on the basis of Genesis 1 in spite of scientific findings;
2. it is based on the disputable assumption that the Bible teaches special creation of each of the various "kinds" of living creatures in a manner that precludes genealogical links;
3. it gives the appearance of saying that the authenticity of the Bible's teaching concerning creation is *dependent* on demonstrating the falsehood of the contemporary scientific world picture;
4. it attempts to demonstrate the falsehood of the contemporary scientific world picture by making capital of
   a. the unfilled gaps in the paleontological record so far recovered,
   b. the inability of scientists up till now to show how life-forms may have emerged from inorganic materials, and

14. See, for example, Alan Hayward, *Creation and Evolution: The Facts and Fallacies* (London: Triangle, 1985), Russell W. Maatman, *The Bible, Natural Science, and Evolution* (Sioux Center, IA: Dordt College Press, 1970), and Pattle P. T. Pun, *Evolution: Nature and Scripture in Conflict?* (Grand Rapids: Zondervan, 1982).

15. See, for example, Jitse M. van der Meer, "Hierarchy: Towards a Framework for the Coherence of Faith and Science," *Pro Rege*, March-June 1989, pp. 19-33.

c. the failure of scientists yet to agree on a theory as to how the more complex life forms may have developed from the more simple,

but such an approach appears especially vulnerable in view of the fact that all of these continue to be areas of intense scientific investigation; and

5. it fails to account for much empirical data already in hand, not least in the newly developed area of molecular genetics.

How will the issue be decided? Only through the extended cooperative efforts of Christian scholars in the sciences, in biblical studies, and in theology and philosophy. Perhaps it will be settled in much the same manner as were questions regarding the motion of the Earth or the timetable of geological history.

HOW, THEN, SHALL WE PROCEED?

Making no pretense of being exhaustive, we offer here a few guidelines for our continuing thought and action with regard to the issues addressed in this book. In view of the diversity of local circumstances, each reader will need to select, adapt, or modify these guidelines as sound judgment dictates.

1. *We may never allow the authenticity of the Christian faith to be held dependent upon the accuracy of one specific picture concerning the historical particulars of God's work in forming the highly diversified contents of his Creation.*

Regardless of whether the recent special creationist picture or the old-earth special creationist picture or an evolutionary creationist picture or some other picture yet to be devised offers the most accurate *description* of God's creative activity, the fundamental Christian *doctrine* of creation holds true: there is only one God—the God who reveals himself to us in the Bible, in history, and in his Son Jesus Christ—and he is the Creator of all else; every other being or thing is creature and is dependent on God for all things.

2. *We may not allow the warfare metaphor to control our perception of the relationship between natural science and Christian faith.*

If, as we Christians confess, the entire universe is God's Creation and if the whole of its formative history is his handiwork, then we should be eager to employ the most powerful tools available for exploring it. The natural sciences, when conducted with competence and professional integrity, provide us with an empirical access to God's handiwork that we may neither neglect nor treat with disdain. Stated more positively, we are called to acknowledge the fruits of honest and competent scientific investigation as we strive to articulate our Christian witness to a scientifically knowledgeable culture. We are called to demonstrate that Christian faith and sound scholarship are not at war with one another—that Christian faith is neither dependent upon ignorance nor threatened by knowledge about the world of which we are a part. By example we must show that Christian faith gives purpose to scientific scholarship and meaning to the knowledge thereby gained.

3. *We may not allow the warfare metaphor to control our perception of the relationship between biblical scholarship and Christian faith.*

Recognizing the sufficient perspicuity of God's revelation in Scripture, the Christian church has long confessed that even the most humble reader of the Bible will be capable of grasping all that is necessary for hearing God's call to repentance, his offer of redemption, his reassurance of fatherly care, and his will for faithful service. Nonetheless, we also recognize that the Bible is an unfathomably rich reserve of God's disclosures in a form that is sometimes obscure and culturally alien to us and that it therefore demands continuing and intensive study by those who are specially called and trained to do so. The results of faith-grounded biblical scholarship ought never to be neglected or treated with disdain but rather ought to be welcomed and incorporated into the articulate expression of our Christian faith in this age. While we may in no way insist that sincere and authentic faith is dependent upon a scholarly grasp of the subtleties of Scripture or theology, neither may we encourage an anti-intellectualism that views a simplistic faith as inherently superior to a faith shaped and informed by Spirit-gifted and Spirit-guided biblical scholarship.

We are especially concerned with the mischievous form of "folk exegesis" that is characterized by at least these two features: (1) a tendency to neglect the relevance of the specific times and circumstances in which God inspired the various portions of Scripture to be written—in effect, to wrench the biblical text out of its historical and cultural context; and (2) the practice of interpreting the Bible by viewing it through the filter of an unexamined assemblage of hermeneutical assumptions and traditions—a practice often accompanied by the notion that such a reading yields "the plain meaning of the text" or constitutes "just taking

God at his word," while denigrating the more scholarly approaches as being "fanciful interpretations," or as being outright rejections of "what God has plainly said."

4. *We may not hold the received tradition as being infallible.*

The heritage bequeathed to us by faithful Christian individuals and ecclesiastical communities of the past is quite correctly held in high respect. The faith of our spiritual parents is not to be minimized or ignored. Yet it must be honestly recognized that this inherited articulation of the faith is in part a human construction and thus must remain open to modification and reformulation. Furthermore, we confess that God continues dynamically to lead his church in these times as well as in the past. We are, in fact, continually called to examine our heritage critically and to rearticulate the historic Christian faith in the contemporary context with its peculiar opportunities, agenda, and vocabulary. In the Reformed heritage, for example, we insist that "to be Reformed is always to be reforming." That's a good motto, but not an easy one to follow. The "all-or-nothing" attitude toward the "tradition packet" is thereby forbidden; we are called instead to be willing always to assess critically our understanding both of Scripture and the created world—not to doubt or disbelieve God's revelation in either but to question *our* understanding of it. That reexamination can be very unsettling. It is our nature to fear uncertainty. But in order to consider any modification we must first be willing to admit a degree of tentativeness about our starting point. Hence we encourage here no change merely for the sake of novelty but change only in response to God's leading.

5. *We need to encourage and promote continuing study in those areas that present difficult and challenging questions.*

When the conceptual framework provided by a sound and vigorous heritage is applied to the questions stimulated by contemporary scholarship, fruitful insights may well be generated. Believing that the heritage of Christian scholarship is sound and that many questions occasioned by modern scientific investigation are important, we are eager for vigorous research and study to proceed.

While we are committed to believing whatever the Bible authoritatively teaches, we do not have an infallible interpretation or understanding of all the biblical passages. Hence biblical scholarship must be encouraged.

As believers we are committed to coming to know God "by the creation, preservation, and government of the universe; which is before our eyes as a most elegant book, wherein all creatures, great and small, are as so many characters leading us to see clearly the invisible things of God, even his everlasting power and divinity" (Belgic Confession, Art. 2). And while natural science does not in itself have the competence to show us God, scientific scholarship can provide us in remarkable ways with helps for perceiving God's revelation to us, and it must be encouraged.

In order to articulate and systematize the contemporary expression of our historic Christian faith, we need theologians and philosophers who can knowledgeably incorporate the fruits of both biblical and scientific scholarship into their theorizing. The efforts of such people must be encouraged.

The necessary context in each case is found in the last word of each of the preceding three paragraphs: *encouraged.* The responsibility lies not only with those who do the scholarship but with each member of the community of Christian encouragers. Together we search for answers to the difficult questions—not because our faith (our trust in and commitment to God) depends upon the answers but because as stewards of the Creation we are charged to seek knowledge and because in faith we have the confidence that this search will draw us toward a heightened appreciation of the awesome majesty and mystery of God and his ways.

INDEX OF PRINCIPAL NAMES

SCRIPTURE REFERENCE INDEX